Mark Ridley works in the Department of Zoology, Oxford University. He previously held posts at Emory University, Atlanta, and at Cambridge University. His books include the standard college text *Evolution*, and *Mendel's Demon*. He has also edited the anthologies *Evolution* and *A Darwin Selection*.

HOW TO READ

HOW TO READ

DARWIN

MARK RIDLEY

W. W. Norton & Company

New York   London

First published in Great Britain by Granta Publications

For information about permission to reproduce selections from this book,
write to Permissions, W. W. Norton & Company, Inc.,
500 Fifth Avenue, New York, NY 10110

Manufacturing by The Maple-Vail Book Manufacturing Group
Production manager: Amanda Morrison

Library of Congress Cataloging-in-Publication Data

Ridley, Mark.
How to read Darwin / Mark Ridley.— 1st American ed.
p. cm. — (How to read)
Includes bibliographical references and index.
ISBN-13: 978-0-393-32881-3 (pbk.)
ISBN-10: 0-393-32881-3 (pbk.)
1. Darwin, Charles, 1809–1882. 2. Evolution (Biology) 3. Natural
selection. I. Title. II. How to read (New York, N.Y.)
QH31.D2R475 2006
576.8'2—dc22
2005035281

W. W. Norton & Company, Inc.
500 Fifth Avenue, New York, N.Y. 10110
www.wwnorton.com

W. W. Norton & Company Ltd.
Castle House, 75/76 Wells Street, London W1T 3QT

1 2 3 4 5 6 7 8 9 0

CONTENTS

SERIES EDITOR'S FOREWORD

How am I to read *How to Read*?

This series is based on a very simple, but novel, idea. Most beginners' guides·to great thinkers and writers offer either potted biographies or condensed summaries of their major works, or perhaps even both. *How to Read*, by contrast, brings the reader face-to-face with the writing itself in the company of an expert guide. Its starting point is that in order to get close to what a writer is all about, you have to get close to the words they actually use and be shown how to read those words.

Every book in the series is, in a way, a masterclass in reading. Each author has selected ten or so short extracts from a writer's work and looks at them in detail as a way of revealing their central ideas and thereby opening doors onto a whole world of thought. Sometimes these extracts are arranged chronologically to give a sense of a thinker's development over time, sometimes not. The books are not merely compilations of a thinker's most famous passages, their 'greatest hits', but rather they offer a series of clues or keys that will enable readers to go on and make discoveries of their own. In addition to the texts and readings, each book provides a short biographical chronology and suggestions for further reading,

Internet resources, and so on. The books in the *How to Read* series don't claim to tell you all you need to know about Freud, Nietzsche and Darwin, or indeed Shakespeare and the Marquis de Sade, but they do offer the best starting point for further exploration.

Unlike the available second-hand versions of the minds that have shaped our intellectual, cultural, religious, political and scientific landscape, *How to Read* offers a refreshing set of first-hand encounters with those minds. Our hope is that these books will, by turns, instruct, intrigue, embolden, encourage and delight.

Simon Critchley
New School for Social Research, New York

INTRODUCTION

How should we read Darwin? He was a historical figure, who revolutionized biology, and his influence can be traced into all corners of modern culture – not just in biology, but in philosophy, the human sciences, theology, software engineering, literature and the plastic arts. It therefore makes sense to consider the relation between what Darwin wrote and the big themes of human thought. Darwin was also a mastermind, and it is fascinating to read him independently of his historical influence. You can read him as an almost private conversation. He thought widely, and almost always had original and stimulating things to say. It is a delight to come into contact with such a mind. You soon come to recognize the way he thought; he liked to gather all the facts he could, from an amazing array of sources. He liked to build an abstract, general theory that made sense of the crucial features of the subject he was working on. He thought through all the difficulties, with a rare intellectual honesty.

Darwin was a prolific writer. He wrote not only one of the most influential books of all time, *The Origin of Species*, but also a large range of other books on topics as diverse as coral reefs, barnacles, earthworms and orchids, as well as a travel book based on the voyage of the *Beagle* and an autobiography. I have allocated most space here to *The Origin of Species*, at the cost of paying less attention than I should have liked to the rest of his work. I have also discussed what is probably his second most famous book, *The Descent of Man*, and sampled the leading idea from *The Expression of the Emotions*.

Should we read him as historians, or as scientists? Darwin the scientist has inspired about as much subsequent research as anyone ever has, and modern science has added a pile of subsequent discoveries to each of Darwin's smallest observations. Readers will, I suspect, want to understand both what Darwin was saying, in relation to what he was thinking about at the time, and whether his theories are still thought to be correct after 150 years of subsequent investigation.

For a 'historical' reading of Darwin, my aim is to get inside Darwin's head – to try to understand how he was thinking. My preferred approach follows the philosopher and historian R. G. Collingwood, who suggested we should aim to understand the questions that people asked. The questions they had in mind may not be explicit, but if you can puzzle them out, the rest of their work tends to make more sense. When faced with a barrage of Darwinian evidence, drawn from agriculture and the obscurer reaches of natural history, we should ask what question Darwin had posed himself. When we see an abstract argument, we should ask what question the argument is intended to answer. In this way, we can come to understand what people in the past were thinking. Collingwood's method forces us to be active readers, looking for sense in the text, rather than reading passively and hoping to absorb something.

Many readers will not want to understand Darwin only in the context of his own times. Darwin is too big to be left to historians alone. He is massively influential today, and I also like to read him from a modern perspective. For instance, Darwin's detailed understanding of heredity is now universally rejected and has been replaced by genetics. Yet Darwin's main ideas have actually been strengthened by what we now know about genetics and DNA, and I suspect he would have been delighted by these advances. It therefore makes more sense for most of us to understand Darwin's evolutionary ideas in genetic terms rather than in terms of Darwin's own theories of inheritance.

For the modern reader, Darwin has one huge advantage over every other great scientific writer and over many other great writers in general. He is not only important; he is also readable. He wrote for the general reader of his time and his writings contain few technicalities, and no mathematics. If non-specialists try to read Copernicus, Newton or Einstein, the chances are they will get little in return for their effort. Likewise, there are many more recent scientists who are undoubtedly important, yet only an expert minority can understand their papers. Darwin is an almost unique exception. He wrote at a time when there was a substantial educated audience who were interested in the big questions of science. Earlier scientists tend to be lost in history. More recent scientists are lost in specialization and technicality. Most of us can come into contact with big scientific ideas only via an educational intermediary or a popularizer of science. Educational and popular writers can be top-rate, but they are never quite the same as going directly to the source: with Darwin's *The Origin of Species* we can do just that.

'ONE LONG ARGUMENT'

The Origin of Species 1

Darwin once called *The Origin of Species* 'one long argument', but it can be broken down into two more manageable parts. One part considers whether modern life forms have arisen by evolution or separate creation. Darwin makes the case for evolution. (However, Darwin himself used the expression 'descent with modification' rather than 'evolution'. The term 'evolution' came to be used soon after Darwin's book was published in 1859.) According to the theory of evolution, the various forms of life on Earth – trees and flowers, worms and whales – all descend from common ancestors. These ancestors looked very different from their modern descendants. The alternative view, which Darwin argues against, is the theory of separate creation, or creationism. According to the theory of separate creation, the ancestors of the modern forms of life looked much like the modern forms, and the various forms of modern life have separate, rather than common, origins. A religious version of the theory of separate creation would also claim that each life form was supernaturally created by God. Darwin argued against the theory of separate creation, but not against religion; he denied that species have separate origins, not the existence of God.

The second part of *The Origin of Species* considers the process that causes evolution. Darwin argues that the process he calls natural selection drives evolution. These two arguments – one for evolution and the other for natural selection – overlap in the book. The earlier chapters of the *Origin*, particularly numbers 3, 4, 6, 7 and 8, are more about natural selection and the later chapters (9–14) more about evolution; but both issues crop up in every chapter of the book.

The first two chapters of *The Origin of Species* discuss heredity and variation. Heredity (also called inheritance) refers to the way offspring in some respects resemble their parents: taller than average parents tend to produce taller than average offspring. The biological mechanisms that underlie heredity have been worked out since Darwin's time. Inheritance is now known to be caused by genes and DNA; but in Darwin's time the mechanism of heredity was an unsolved problem. Variation refers to differences between individuals within a population (or within a sample). As the extract below shows, Darwin tends to say 'individual variability' where a modern biologist would say 'variation'. Either way, these terms are used in a non-temporal sense, to refer to the various forms that exist within a species. The human species in this sense shows variation with respect to size, personality, skin colour, and so on. In colloquial language, this is often called diversity, and words derived from the verb 'vary' often refer to change over time. Biologists, however, have come to use 'variation' to refer to inter-individual differences at any one time, and they tend to use 'diversity' to refer to difference between species. 'Biological diversity' refers to the full array of life, from microbes to coral reefs to tropical forests.

Darwin began his book with heredity and variation because his whole theory depended on them. Darwin's theory needed inheritance: if new (that is, variant) forms of a species are not inherited, evolution will not occur and natural selection will not work. Darwin demonstrated the existence of heredity and variation using evidence from agricultural varieties and from pigeon breeding. This material is not the best way to introduce the theory to modern readers. Modern authors would write about genetics instead. The biggest difference between the way we understand evolution now, and Darwin's own understanding, lies in inheritance. At an abstract level, Darwin's argument is watertight. All he really needed to show was that heredity occurs somehow and that variation exists. However, his detailed material on the two topics is no longer current.

Darwin introduces the theory of natural selection in chapters 3 and 4. He begins by connecting his earlier material on heredity and variation to the theory he is about to discuss and then gives an advance overview of the theory of natural selection itself.

Before entering on the subject of this chapter, I must make a few preliminary remarks, to show how the struggle for existence bears on Natural Selection. It has been seen in the last chapter that amongst organic beings in a state of nature there is some individual variability; indeed I am not aware that this has ever been disputed. It is immaterial for us whether a multitude of doubtful forms be called species or sub-species or varieties; what rank, for instance, the two or three hundred doubtful forms of British plants are entitled to hold, if the existence of any well-marked varieties be admitted. But the mere existence of individual variability and of some few well-marked varieties, though necessary as the foundation for the work, helps us but little in understanding how species arise in nature. How have all those exquisite adaptations of one part of the organisation to another part, and to the conditions of life, and of one distinct organic being to another being, been perfected? We see these beautiful co-adaptations most plainly in the woodpecker and misseltoe; and only a little less plainly in the humblest parasite which clings to the hairs of a quadruped or feathers of a bird; in the structure of the beetle which dives through the water; in the plumed seed which is wafted by the gentlest breeze; in short, we see beautiful adaptations everywhere and in every part of the organic world.

Again, it may be asked, how is it that varieties, which I have called incipient species, become ultimately converted into good and distinct species, which in most cases obviously differ from each other far more than do the varieties of the same species? How do those groups of species, which constitute what are called distinct genera, and which differ from each other more than do the species of the same genus, arise? All these results, as we shall more fully see in the next chapter, follow inevitably from the struggle for life. Owing to this struggle for life, any variation, however slight and from whatever cause proceeding, if it be in any degree profitable to an individual of any species, in its infinitely complex relations to other organic beings and to external nature, will tend to the preservation of that individual, and will generally be inherited by its offspring. The offspring, also, will thus have a better chance of surviving, for, of the many individuals of any species which are periodically born, but a small number can survive. I have called this principle, by which each slight variation, if useful, is preserved, by the term of Natural Selection, in order to

mark its relation to man's power of selection. We have seen
that man by selection can certainly produce great results,
and can adapt organic beings to his own uses, through the
accumulation of slight but useful variations, given to him
by the hand of Nature. But Natural Selection, as we shall
hereafter see, is a power incessantly ready for action, and
is as immeasurably superior to man's feeble efforts, as the
works of Nature are to those of Art.

The quotation begins with the distinction between what I
called the two main arguments of the *Origin* – between evolution and natural selection. It then moves on to an argument
about variation, or 'individual variability', and its relation
with higher classificatory groups..

Biologists classify living things into a hierarchy of groups.
The biggest groups are things like 'animals' and 'plants'. Then
at successively lower levels are groups such as vertebrates,
mammals, primates, apes and human beings. The lowest level
group is usually the species (human beings are a species).
However, Darwin mentions two even lower levels, subspecies
and varieties. These are not clearly different kinds of category;
both refer to some distinct recognizable grouping within a
species. He uses the word 'variety' more. A variety is something like a breed of dogs (poodles, terriers and so on) or a
geographic race. Above the species level, the next immediate
level is the genus. For example, humans are in the genus
Homo, which includes both us and some extinct but closely
related species.

Individual variability – the kinds of difference you see
between two individual organisms – is variation on the smallest scale. 'Varieties' show larger differences: one terrier will
differ from another terrier in certain details, but terriers are

distinct from St Bernards. Variation mattered to Darwin because it undermined the theory of separate creation. People who think that species are separately created tend also to think that each species is a distinct form of life, recognizably different from other forms of life. However, different varieties (of pigeons, for example) show a range of degrees of difference. In some cases, two varieties may be similar, in others they are somewhat different, and in others they differ more than do two conventionally classified species. Thus, the impression that each species is distinct is naive. When you look carefully, you find that variation within a species blurs into differences between species. If, as a creationist, you tried to specify which entities were separately created, you'd soon get hopelessly confused. Is it species, or varieties, or individual variations, that are separately created? Any answer will seem arbitrary, because the degrees of difference overlap. Living things do not exist in clearly distinct forms.

Variation also matters in the theory of natural selection. In the extract, Darwin sets out the problem in general terms, in a striking way. He asks two questions: what is the explanation for adaptation, and what is the explanation for continual evolutionary change. These two questions provide two tests for any proposed theory about what drives evolution. If the theory cannot explain adaptation and continual evolutionary change, the theory is inadequate.

Adaptation is a basic problem in biology. Darwin (and modern biologists) again uses the word in a specialist, technical sense that differs slightly from its colloquial usage. In colloquial use, adaptation usually refers to change over time. We might talk about someone 'adapting' to a new job – that is, how they adjust their behaviour to the new conditions. When Darwin talks about 'all those exquisite adaptations' he is referring to structures such as hands and eyes, that exist in a form that is well designed, given the life that the creature is leading. Eyes, for instance, have an optical structure, with a lens and light-sensitive cells, that enables them to be used for vision. An eye is an example of an adaptation. An adaptation is any part of a body (or its behaviour) that is well designed for life.

Adaptation is a special, highly non-random state of nature that would not just arise automatically or by chance. It requires explanation. Before Darwin, many people explained it by the supernatural action of God. Indeed the existence of adaptation in nature provided one of the main philosophical arguments for the existence of God, called the argument from design. Darwin's theory of natural selection made it unnecessary to posit the existence of God, at least to explain adaptation in nature.

Natural selection does successfully explain adaptation, as the brief overview in the above quote indicates (and as we shall see further in Chapters 2 and 3). However, many other theories have been put forward to explain evolution and most of them do not successfully pass this test. For instance, some biologists since Darwin have proposed that evolution proceeds in big jumps, driven by special kinds of rare, large genetic change (sometimes called macro-mutations). Changes in the DNA have no particular tendency to produce adaptation; they are as likely to be changes for the worse as for the better. (Indeed, in creatures that are already well adapted, a large change is most likely to be a change for the worse.) Macro-mutational theories of evolution are unable to explain adaptations and, as far as Darwin is concerned, fall at the first hurdle.

In some respects, modern biologists place less emphasis on

Darwin's first test – whether the theory explains adaptation – than did Darwin. They also place more emphasis on random evolutionary change than he did. Two main processes are now accepted to cause evolutionary change: natural selection and random genetic drift. That is, evolution can not only be driven by selection, as Darwin argued; it can also happen by chance if there are two equally good versions of a gene (or of a stretch of DNA) and one is luckier than the other during reproduction over the generations. The new emphasis on random evolutionary change is a consequence of the discovery of DNA. Darwin only knew about the observable form of organisms, and was concerned with evolution in these features. Almost all the large-scale, observable properties of organisms are adaptations. As such, they almost certainly evolved by natural selection. Random genetic drift cannot drive adaptive evolution – because adaptation is almost by definition non-random. However, it has turned out that adaptive evolution accounts for only a fraction of evolutionary change in the DNA. Perhaps as little as 5% of the DNA in a human being actually codes for the body. The other 95% may (though this is uncertain) mainly be 'junk DNA': DNA that is reproduced from parent to offspring and does no harm but is more or less useless. Evolution in this junk DNA is non-adaptive and random. It could not be adaptive, because the junk DNA does not code for anything in the body.

The changed emphasis on random evolution, between Darwin and now, follows from the way we now tend to think of evolution in terms of changes in the DNA. Darwin would probably agree that most evolution is driven by random processes, rather than natural selection, if he knew what we now know about DNA. For example, the DNA sequences of human beings and mice are now almost completely transcribed. Of the 3,000 million or so letters of human DNA, about a sixth, or 500 million, have changed in the 100 million years or so since our common ancestor with mice was creeping round in the shadow of the dinosaurs. It may be that in round figures only about 25 million of those changes, or fewer, were needed to change that ancestral mammal into a human. Most of those 25 million would have been driven by natural selection. Against that, more like 475 million changes have occurred by random evolution. Natural selection still explains why our bodies have evolved to be well designed, but random evolutionary processes now have an unavoidable quantitative claim on our attention. That was not so when Darwin was writing.

Darwin's second question, and second test of a proposed theory of evolution, is whether it explains evolutionary change. In particular, any theory of evolution has to be able to account for the full diversity of life. A theory is inadequate if it only explains small-scale evolution, or evolution in a different pattern from that seen in life on Earth. Biological diversity has a hierarchical pattern, reflected in its hierarchical classification into species, genus and so on. The reason for this pattern is presumably that different forms tend to evolve apart, or diverge, over time. Different forms with a recent common ancestor are still relatively similar; but those with a more distant common ancestor have become more different. Darwin was therefore on the look-out for a theory in which different forms evolved apart, and perhaps even pushed each other apart, over time. This search underlies many of Darwin's remarks as he discusses the 'struggle for existence'. Darwin was almost unique in understanding that competition is strongest between individuals within a population, rather than occurring between races, or species. As we shall see in the next chapter, this led him to his 'principle of divergence' that enabled him to answer his second question about evolution.

At the end of the quoted passage, it is worth noticing how Darwin points to the relation between 'selection' in nature and 'selection' by humans. Humans breed agricultural and domestic varieties (such as pigeons), by selectively breeding from individuals with a high milk yield or fancy set of feathers. Darwin was an expert on the topic, and repeatedly used an analogy with human 'artificial' selection when thinking about natural selection. Modern writers rarely introduce the theory of natural selection in this way, probably because it has become less familiar. But the reader of Darwin is soon transported into a world where science and agriculture form a confident, not a controversial, partnership.

In Darwin's time, his theories of evolution and of natural selection received a variety of responses. The idea of evolution − that species change over time − had been proposed many times before. It exists in the writings of ancient Greece, and had been criticized, discussed and revived many times since. Many, perhaps most, biologists found Darwin's case for evolution compelling. In the early and mid-nineteenth century, few biologists openly accepted evolution; in the late nineteenth century only a minority of them openly opposed the idea: this change was mainly due to Darwin. The particular form that evolution took in Darwin's theory did differ from other evolutionary theories, but these differences were details; evolution itself had become part of the biological mainstream.

Natural selection was an altogether more original idea. Some rudimentary traces of it do exist before Darwin, but they had been scarcely thought through and had practically no influence. Darwin saw how natural selection could act as a creative force and explain essentially all the evolution of life; no one had remotely had that idea before. Few people understood natural selection when Darwin put it forward, and it was almost universally rejected or ignored. Biologists only came to take the theory of natural selection seriously, by degrees, in the first half of the twentieth century. It was not widely accepted, as the explanation for evolution, until about 1950.

There is one interesting exception. Darwin thought up natural selection in the late 1830s, but then kept quiet about it. He was preparing to write a big book on the subject, and took his time. Then, in 1857, he received a letter from Alfred Russel Wallace, who had thought up almost the same idea. Wallace (1823–1913) was another British naturalist who, like Darwin, travelled the globe. Wallace wrote to Darwin from Malaya. His letter prodded Darwin into action, and the theory of evolution by natural selection was first made public in a joint publication by Darwin and Wallace in 1858. That paper was ignored. Meanwhile, Darwin started to write up the theory in book-length, in what he called an abstract of the 'big species book' he had been at work on. That abstract was *The Origin of Species* and, far from being ignored, caused a sensation. Wallace was always generous in giving credit to Darwin as the main author and creator of the theory of evolution by natural selection; but it is worth remembering that, for all Darwin's originality, someone else had had much the same idea soon after him.

2

NATURAL SELECTION

The Origin of Species 2

A struggle for existence inevitably follows from the high rate at which all organic beings tend to increase. Every being, which during its natural lifetime produces several eggs or seeds, must suffer destruction during some period of its life, and during some season or occasional year, otherwise, on the principle of geometrical increase, its numbers would quickly become so inordinately great that no country could support the product. Hence, as more individuals are produced than can possibly survive, there must in every case be a struggle for existence, either one individual with another of the same species, or with the individuals of distinct species, or with the physical conditions of life. It is the doctrine of Malthus applied with manifold force to the whole animal and vegetable kingdoms; for in this case there can be no artificial increase of food, and no prudential restraint from marriage. Although some species may be now increasing, more or less rapidly, in numbers, all cannot do so, for the world would not hold them.

There is no exception to the rule that every organic

being naturally increases at so high a rate that if not destroyed, the earth would soon be covered by the progeny of a single pair. [. . .]

A corollary of the highest importance may be deduced from the foregoing remarks, namely, that the structure of every organic being is related, in the most essential yet often hidden manner, to that of all other organic beings, with which it comes into competition for food or residence, or from which it has to escape, or on which it preys. This is obvious in the structure of the teeth and talons of the tiger; and in that of the legs and claws of the parasite which clings to the hair on the tiger's body. But in the beautifully plumed seed of the dandelion, and in the flattened and fringed legs of the water-beetle, the relation seems at first confined to the elements of air and water. Yet the advantage of plumed seeds no doubt stands in the closest relation to the land being already thickly clothed by other plants; so that the seeds may be widely distributed and fall on unoccupied ground. In the water-beetle, the structure of its legs, so well adapted for diving, allows it to compete with other aquatic insects, to hunt for its own prey, and to escape serving as prey to other animals.

The store of nutriment laid up within the seeds of many plants seems at first sight to have no sort of relation to other plants. But from the strong growth of young plants produced from such seeds (as peas and beans), when sown in the midst of long grass, I suspect that the chief use of the nutriment in the seed is to favour the growth of the young seedling, whilst struggling with other plants growing vigorously all around. [. . .]

Divergence of Character. The principle, which I have designated by this term, is of high importance on my theory, and explains, as I believe, several important facts. [. . .] I believe it can and does apply most efficiently, from the simple circumstance that the more diversified the descendants from any one species become in structure, constitution, and habits, by so much will they be better enabled to seize on many and widely diversified places in the polity of nature, and so be enabled to increase in numbers. [. . .] . . . it should be remembered that the competition will generally be most severe between those forms which are most nearly related to each other in habits, constitution, and structure. Hence all the intermediate forms between the earlier and later states, that is between the less and more improved state of a species, as well as the original parent-species itself, will generally tend to become extinct.

Natural selection operates in any population that meets a number of conditions. One is that some members of a population differ from other members of the population: that is, the population shows *variation*. A second is that offspring tend to resemble their parents: that is, there is *inheritance*. And the third is that some types of individual in the population produce more offspring than average. In a population that meets these conditions, the next generation will contain more of the types of individual that were reproductively successful in the previous generation. Natural selection has driven evolutionary change, and in the direction of improved adaptation; that is, the reproductively successful types of individuals will be the ones that are best adapted to the local conditions.

The argument so far is logically coherent, but incomplete. Darwin needed to show not only that natural selection can act but also that it is so ubiquitous and powerful that it can explain every adaptation in every living creature, and also explain the full diversity of life. After all, a critic might accept that natural selection can act, but deny its importance – maintaining that it explains only a fraction of adaptive evolution. Some critics still argue just that.

Darwin's argument for the ubiquitous, powerful operation of natural selection is ecological. It refers to the relation between organisms and resources and competitors in the external environments. (The word 'ecology' – the science of the relation between living things and their environments – did not exist at the time. It was coined by Darwin's German follower, Ernst Haeckel, in 1873, but did not become widely used until the mid-twentieth century.) Darwin is extending 'the doctrine of Malthus' to all life. Thomas Robert Malthus (1766–1834) had published his *Essay on the Principle of Population* in several editions from 1798 to 1830. Malthus argued that the human population would tend to increase at a higher rate than the food supply. Malthus's *Essay* was widely read in Darwin's time, and Darwin himself recalled in his autobiography how he happened to pick it up as general reading in the late 1830s. It was this reading that inspired Darwin, leading to his invention of the theory of natural selection.

Darwin realized that not just humans, but all life forms, tend to produce far more offspring than can be supported by the food supply. The result is a competition to survive, or what Darwin called the struggle for existence. He explained that the word 'struggle' is a metaphor, and often does not take the form of a physical struggle or fight. In every species, only a minority of the offspring survive each generation. Just what competitive attributes an individual offspring needs in order to survive will depend on the way of life in that species. The form of competition differs between bacteria in our guts, corals on a reef, and monkeys in a jungle. Darwin clarifies his meaning of the word struggle by considering how mistletoe seeds might effectively be competing against each other, and the seeds of other plants, to be picked up and spread by birds. Clearly there is no physical fight between the seeds, but there is a metaphorical struggle.

A double purpose lies behind Darwin's chapter on the struggle for existence. He wants to persuade the reader that the competition for survival leads to natural selection. This in turn explains why we see adaptive design in living things, and why there is evolutionary change. However, he also wants to persuade us that the struggle is universal and intense, notwithstanding the apparent peacefulness of nature. In a famous metaphor, he said, 'The face of Nature may be compared to a yielding surface, with ten thousand sharp wedges packed close together and driven inwards by incessant blows, sometimes one wedge being struck and then another with greater force.' The 'wedges' here are the reproducing, competing units of nature. The economy of nature is packed tight. The continual reproduction of organisms and their competitors means that the packing tends to become ever tighter.

Darwin was also keenly aware of the web of ecological relations among species. His best-known analysis of the kind concerned the relation of clover, bumblebees (which pollinate clover, and which Darwin called humblebees), mice (which eat bumblebee nests), and cats. If more cat-lovers moved into an area, we might expect it to have an effect on the local clover population. In Darwin's view of ecology, each species produces far more offspring than can survive, resulting in intense competition at least within each species – and to a lesser extent between different species too. Also, each species is connected to many other species, which can lead (via the webs of relationship) to subtle and surprising pressures. In this way, Darwin is able to offer an explanation for the details of adaptive design in each species.

If you have a simple view of the struggle of existence you can understand the teeth and talons of a tiger. But you need to take seriously the full range of ecological relations of each species to appreciate why it has its subtler adaptive features. The size of the seed, for instance, reflects the amount of food it has been supplied with by the parent plant. A better supplied seed can grow faster, and out-compete other seeds in the struggle up to the light and in the limited growing space of an already packed field. Ecological relationships, and the many forms of competition, are for Darwin the key to understanding all the adaptive design in nature.

Darwin's view of ecology was new in his time. None of his contemporaries wrote about the topic in the same way, not least because most biologists had an educational background in medicine. They were more familiar with the anatomy of dead bones than with live creatures. They did not see each species as part of a web of relations with other species and forces in the natural environment. Darwin saw how each species was adapted for interactions with natural enemies such as parasites, with competitors and with its resource supplies. This Darwinian view has now become orthodox. Nature programmes on TV typically come with a more or less Darwinian commentary. Academic ecology traces itself back to chapter 3 of *The Origin of Species.* The authors of one standard modern college text of ecology only semi-humorously said they wanted to reprint Darwin's chapter verbatim as chapter 1 of their book. If Darwin had not invented the theory of evolution, he would still be remembered as the inventor of ecology.

Modern biologists also follow Darwin in using ecological competition to understand the forms of adaptation in nature. Indeed the main trend, in scientific research on adaptation, has been to deepen and extend Darwin's style of reasoning rather than seriously revise it. For example, there is a phase of competition even before the growth of young plants from seeds which Darwin discussed. Before seeds are produced, the eggs must be fertilized. In plants, this occurs after pollen are deposited on the female parts of the plant. Each pollen then develops what is called a 'pollen tube' that grows towards the egg. When the tube reaches the egg, it can transfer male DNA from the pollen and fertilization can occur. However, things are not so simple. Any one plant may receive pollen from more than one other plant. Several pollen grains may start to grow pollen tubes towards the same egg. The result is competition between pollen, before the seeds are even produced. Pollen have evolved adaptations to grow their tubes rapidly, and perhaps even such 'dirty tricks' as an ability to inhibit the growth of other pollen tubes.

Before the pollen reach their destination, they have to be picked up by a pollinating insect (if we ignore plants that use wind or water to transfer pollen). Plants compete for the attention of insects by developing colourful flowers, with nectar as the reward for the visiting insects. Biologists now see flowers as adaptations somewhat analogous to the colourful feather designs of male birds such as peacocks. Chapter 9 in this book discusses Darwin's ideas on the topic.

We saw how Darwin began his discussion of natural selection with two questions, which corresponded to two tests for any proposed theory of evolution: does it explain adaptation? And does it explain the 'tree-like' diversity of life? Darwin's argument about adaptation, which I have been considering so far, is probably the more familiar of the two. Modern biologists still think in much the same way as Darwin did about adaptation and ecological competition. Darwin's argument about diversity may be less familiar. The argument, which Darwin called the 'principle of divergence', was clearly important to him. He devotes as much space to it in the *Origin* as he does to adaptation. Darwin had immediately realized how natural selection would explain adaptation when he first thought of natural selection in the late 1830s. He wrote two essays about evolution by natural selection in the early 1840s, and though they were not published, the manuscripts survive. We can see that Darwin had in 1840 almost the same theory as was finally published in the *Origin* in 1859. The one big difference is that the essays of the 1840s lack the principle of divergence. In his autobiography, he describes how his development of the principle plugged what he saw as the main hole in his theory. Thus it is impossible to understand *The Origin of Species* without understanding the principle of divergence.

Darwin's question was why has evolution proceeded divergently — why is it that branches split from time to time, producing two branches that then evolve apart over time? This divergence is reflected in biological classification. Classifications of living things have a form invented by the eighteenth-century Swedish naturalist who signed himself with the Latinized name Carolus Linnaeus. Linnaean classifications are hierarchical; in Darwin's words they have 'groups within groups': several species are classified in a genus, several genera in a family, and so on. This pattern emerges if evolutionary lineages diverge over time. The species within a genus share a recent common ancestor and have diverged only a little. The genera within a family share a more distant common ancestor and have diverged further. Darwin's question was: What is forcing this general pattern of divergence? Why don't species just evolve apart a certain amount, and then stay there? Or why don't they evolve back to resemble one another more after a certain time?

The hierarchy of life – that is, the Linnaean pattern of subgroups within larger groups – was a big theme in biology in the early nineteenth century. Part of Darwin's obsession with explaining it probably reflects the fact that it was a hotly debated topic at the time. Some modern biologists think that the divergent, tree-like pattern of evolution is almost inevitable. If all life on Earth shares a single common ancestor, then almost any kind of evolution after that will end up being approximately divergent and tree-like. There are certain exceptions, which Darwin did not know about. Completely different evolutionary branches may join together, as if two separate branches in the upper part of a tree grew together and then grew up as a single branch. At certain points in evolution, two life forms have merged into one. All (or almost all) the cells in a human body contain DNA in two cellular subcomponents. Most of the DNA is in the cell nucleus. But some DNA is also found in other structures within each cell called mitochondria. The reason is that mitochondria are descended from once free-living bacteria that, about 2,000 million years ago, invaded or were engulfed by other cells. The result was a sort of hybrid cell-within-a-cell, and human beings together with all animal life are descended from that merger event. Thus evolution is not always divergent – but it usually is, and it is worth understanding Darwin's argument about why it is.

Darwin's principle of divergence concerns the relative strength of competition. In a bird such as a robin, for instance, we can consider the relative strength of competition from other robins and from more distantly related creatures such as lizards, fish, insects or plants. In general we can consider the relative strength of competition that any individual experiences from other individuals like itself and from other, different life forms. The competition from the other life forms like itself will generally be much stronger: they will be exploiting similar resources. One bird may compete with another, for seeds or insects as food, or for nest sites. But a bird does not compete with plants for sunlight, or with large carnivores for prey. In a crowded environment, the way to avoid competition is to evolve to become different from other, similar life forms to yourself. Darwin suggested that competition between varieties within a species would cause them to diverge further apart, until they became different species. Then competition between the two species would drive them apart until they became two genera. The principle of divergence drives all evolutionary lineages apart and results in the tree-like pattern of evolution on the grand scale.

Darwin had a distinctive way of thinking about competition. He thought of competition as mainly an intraspecific process, occurring between individuals within a population. Some of Darwin's contemporaries did think about competition, and how it affects life, but they tended to think of competition as between species, or perhaps between races within a species. They mainly lacked Darwin's concept, in which individuals compete to reproduce into the next generation. Indeed Alfred Russel Wallace, the co-discoverer of natural selection, seems to have thought of competition as between varieties, rather than between individuals. This matters, because many features of life cannot be understood except in terms of competition between individuals within a population. In Chapter 9, we shall see how Darwin took the concept of competition within a species further, in his theory of sexual selection.

The section about the principle of divergence is very closely concerned with the origin of species. During the twentieth century a tradition grew up of remarking, in a slightly joking way, that the origin of species was the one topic in evolution that Darwin did not discuss in his book – making the title a bit misleading. But Darwin is very clearly concerned with the topic here. He asks, 'How, then, does the lesser difference between varieties become augmented into the greater difference between species?' He could hardly be more direct than that. However, Darwin's answer, in terms of competition within a species, is not the answer that most biologists in the late twentieth century would give. We shall look some more at the matter in Chapter 4. The critics who have suggested that Darwin was silent about the origin of species have probably overlooked the discussion because it is so different from their own ideas.

What, finally, do biologists today think of the principle of divergence? The question is hard to answer. Unlike adaptation, which is still a research topic among biologists, the principle of divergence has not been persistently discussed, investigated and criticized. Individual biologists have from time to time 'rediscovered' it and written about it, but there has been no widespread, persistent interest in it in the biological community. Biologists have not striven to explain why evolution, on the grand scale, has a tree-like structure. If you asked around, probably many biologists would accept that Darwin's suggested answer is as good an explanation as any, but other processes may be contributing too. And, in so far as the principle of divergence aims to explain how new species evolve, it is no longer the orthodox explanation.

3

DIFFICULTIES ON THEORY

The Origin of Species 3

Organs of extreme perfection and complication. To suppose that the eye, with all its inimitable contrivances for adjusting the focus to different distances, for admitting different amounts of light, and for the correction of spherical and chromatic aberration, could have been formed by natural selection, seems, I freely confess, absurd in the highest possible degree. Yet reason tells me, that if numerous gradations from a perfect and complex eye to one very imperfect and simple, each grade being useful to its possessor, can be shown to exist; if further, the eye does vary ever so slightly, and the variations be inherited, which is certainly the case; and if any variation or modification in the organ be ever useful to an animal under changing conditions of life, then the difficulty of believing that a perfect and complex eye could be formed by natural selection, though insuperable by our imagination, can hardly be considered real. [. . .]

In looking for the gradations by which an organ in any species has been perfected, we ought to look exclusively to its lineal ancestors; but this is scarcely ever possible, and we are forced in each case to look to species of the same group, that is to the collateral descendants from the same original parent-form, in order to see what gradations are possible, and for the chance of some gradations having been transmitted from the earlier stages of descent, in an unaltered or little altered condition. Amongst existing Vertebrata, we find but a small amount of gradation in the structure of the eye, and from fossil species we can learn nothing on this head. In this great class we should probably have to descend far beneath the lowest known fossiliferous stratum to discover the earlier stages, by which the eye has been perfected.

In the Articulata we can commence a series with an optic nerve merely coated with pigment, and without any other mechanism; and from this low stage, numerous gradations of structure, branching off in two fundamentally different lines, can be shown to exist, until we reach a moderately high stage of perfection [. . .] there is much graduated diversity in the eyes of living crustaceans, and bearing in mind how small the number of living animals is in proportion to those which have become extinct, I can see no very great difficulty (not more than in the case of many other structures) in believing that natural selection has converted the simple apparatus of an optic nerve merely coated with pigment and invested by transparent membrane, into an optical instrument as perfect as is possessed by any member of the great Articulate class. [. . .]

It is scarcely possible to avoid comparing the eye to a telescope. We know that this instrument has been perfected by the long-continued efforts of the highest human intellects; and we naturally infer that the eye has been formed by a somewhat analogous process. But may not

this inference be presumptuous? Have we any right to
assume that the Creator works by intellectual powers like
those of man?

The quoted passage comes from a chapter entitled
'Difficulties on Theory'. Earlier chapters had made the positive case for the theory of evolution by natural selection. Now
Darwin turns to discuss what he sees as the main objections to
the theory. It is characteristic of Darwin that he took seriously
all the objections – or all the ones he knew about or could
think of – to each of his theories. He did not argue like a
lawyer, seeking to belittle, or ignore, or divert attention from
objections. He brought the objections into the light and
examined them carefully. Indeed Darwin's own writings have
been a favoured seam for anti-Darwinian writers ever since;
they soon find that Darwin had assembled much of the case
against his theory as well as for it. Unlike Darwin, they ignore
the latter material. Darwin was constructing an argument
from uncertain, incomplete material; his was an exercise in
inference. It made sense for him to test how robust his inference was, by looking at the case against.

The difficulties discussed by Darwin above are still a stock
argument among creationist critics. How can Darwin's theory
explain 'organs of extreme perfection and complication' of
which the standard example is the eye? The eye posed both a
general problem for any theory of biology in Darwin's time,
and also a specific problem for Darwin's own theory. Darwin
was well aware of both issues, and they underlie his discussion.

The general problem was the 'argument from design' – an
argument, from observation of nature, for the existence of
God. The argument dates back to Plato, and had been end-
lessly formulated and reformulated by medieval Christian

philosophers. It was well known in Darwin's time in the version of the British theological philosopher William Paley (1743–1805). Paley's books were set texts at Cambridge University, where Darwin studied, and Paley's argument from design runs as follows. If we find some complex machine, such as a watch, we can infer that someone – a watchmaker – must have made it. A watch is a highly improbable state of nature: it would not somehow spontaneously have come into existence from the ordinary forces of nature working on the elements of nature. You could tell, by looking at its parts, that it has been designed with a purpose; indeed you could tell this, from the arrangement of cogs and springs, even if you did not know what its particular purpose was. Just as we can infer the existence of a machine designer from complex machines, so we can reason from complex organism in nature to the existence of God.

The argument from design was an important (if not essential or universal) prop for religion in Darwin's time, particularly in England. As a Protestant country, the established religion, Anglicanism, could not justify itself by the continuity of its church and churchmen back to St Peter – that argument made more sense in Roman Catholic countries. Protestant churches could try to justify themselves by 'fundamentalist' arguments from the exact wording (suitably interpreted) of the Bible. One strand of Anglican thinking did tend to fundamentalism in the sixteenth and seventeenth centuries, but it became associated with republicanism and fell into disuse after the Restoration of the Monarchy in 1660. A new justification grew up, often called 'rational religion'. That is, reasonable arguments, from observation of nature, would lead you to accept Christianity. Paley's argument is an example.

Darwin was not the first to criticize the argument from design. Philosophers such as Hume and Kant had done so in the previous century, although they could only provide in-principle objections. They pointed out that the argument from design was incomplete. It assumed that there is no natural process that can produce organs like eyes. Such a process, however, could in principle exist and then the analogy between watch and eye would collapse. This objection gained much more force when Darwin described in detail just such a natural mechanism that produced eyes. 'Rational religion' did not survive Darwin.

Darwin's writings are full of allusions to, and criticisms of, the argument from design. He never makes the argument itself explicit, because it was part of the culture of his readers. However, modern readers may not have met the argument from design, or any of 'rational religion', in their education. It needs to be uncovered as one of the concealed questions that Darwin is thinking about. As the quoted passage ends, Darwin is moving into an exact reformulation of the argument from design, in terms of evolution by natural selection. He starts the final paragraph in the style of Paley, comparing the eye to a human-designed machine, a telescope. Telescopes exist because someone thought them out, as a working whole, before building one. The next step, in Paley's argument, is to suppose God did the same with the eye. 'But how do we know God worked like that?' asks Darwin. He goes on to describe the creation process in the theory of evolution by natural selection. To begin with, the parts of the proto-eye, such as the light-sensitive tissues, varied in thickness from one individual to another. Individuals with superior versions left more offspring, and that version increased in frequency. The eye was built up, bit by bit, over millions of years. No one needed to think up the design before the eye was built. The in-principle argument of Hume and Kant had now been fleshed out in minute detail. The 'argument from design' was a mistake.

Another (not unrelated) reason why Darwin chose to discuss the eye was that it might seem to be the sort of structure that cannot evolve by natural selection. In Darwin's theory, a new structure evolves in many small steps, and each step has to be advantageous. Within an eye, a change in one part might seem to require the right changes in the other parts before it can be advantageous. If the lens changes shape, the retina, and the muscles around the lens, must also change position. If the aperture changes size, changes will also be needed in the light-sensitive cells of the retina. And so on. Any one change always seems to require other associated changes.

However, an organ cannot evolve by natural selection if it requires several, independent changes to occur at the same time. Each change, such as in lens shape, is rare when it first arises; it will only be found in one or two individuals. Those individuals are unlikely also to have the right kind of variation in a second, independent property of the eye, such as the distance between lens and retina. A change in lens shape, without the necessary associated changes, is not advantageous. Natural selection will not favour it. For this reason, it might seem that the eye cannot evolve by natural selection. The problem is the apparent need for simultaneous, coordinated changes in several parts. Such is the property that makes the eye a case study in what Darwin calls 'difficulties on theory'. Darwin's reply is that, if you think the problem through, you find that the eye could in fact have evolved in many small stages. It is unnecessary for all (or many of) the eye's components to change simultaneously in the right way. Darwin looked at the eyes of a range of different species. Some species have little more than a light-sensitive spot, others have a pin-hole camera eye (that is, they can form an image but lack a lens), others have an eye with a lens. The full range of eye forms found in different species is large and shows many possible stages that an eye such as that found now in humans could have passed through during evolution.

A comparative argument such as this does not claim to identify the exact sequence of ancestral stages that an organ such as the eye passed through during evolution. The Articulata that Darwin mentions are not ancestors of humans or any other vertebrata. 'Articulata' is not a term much used now. It refers more or less to what we call arthropods – insects, crustaceans (such as crabs and shrimps), and arachnids (such as spiders). They have a hard external surface that acts as an 'exoskeleton'. They have no internal bones. Vertebrata consist of fish, amphibians, reptiles, birds and mammals. They have an internal skeleton of bones and the external surface is usually soft. In general, if we look at the eyes of any modern life form, we are not looking at an eye that is ancestral to ours. We do not have direct evidence of ancestral human eyes. Eyes are soft parts and are not preserved in fossils. Even if we had a series of fossils of increasingly distant human ancestors, we could not study eyes in them. Instead the comparative evidence provides more of an in-principle argument. It shows that the complex human eye could have evolved in stages from simple beginnings, by providing evidence of the stages. A critic, who said that nothing simpler than a human eye could exist and be advantageous to its bearer, would be refuted by the evidence of simple eyes that do exist and are advantageous. A light-sensitive patch of skin is a less advanced optical device than an eye that can form an image, but it can still provide information that may be a matter of life or death for its bearer.

Another way we can show that a complex organ could evolve in small stages is more theoretical. We may lack comparative evidence of the organ in other life forms. Instead we can imagine a series of stages through which the organ could have evolved, with each stage being advantageous. A formal version of this argument uses an engineering model. Even if we lacked comparative evidence about simpler eyes, we could imagine that the simplest beginnings of the eye consisted of light-sensitive cells. We might then reason that it would be advantageous to improve visual acuity (that is, the ability to distinguish between two objects). In the early stages of eye evolution, the creature might be able to distinguish crudely between objects coming from different sides, not between two closer objects. The light-sensitive cells can then be rearranged such that acuity improves. If the cells are invaginated into a U-shaped pit, then it is easier to distinguish the direction of objects around our animal. Thus a change from a flat light-sensitive surface to a pit of light-sensitive cells can be favoured by natural selection. In the past, random mutations in a creature with a flat light-sensitive region of cells might have produced a range of forms – some with large regions of light-sensitive cells, some with smaller regions, some with a slight invagination of the cells, and others with the cells stuck out, and so on. Of all these random changes, natural selection would establish the version with the best arrangement of light-sensitive cells.

The sequence of changes can be studied theoretically because we can build an engineering model that measures the visual acuity of the eyes and proto-eyes of all possible shapes. We can then see whether it is possible, at each stage, to find a small change that improves visual acuity. Darwin hints at this kind of theoretical argument in the opening part of the quoted passage, but it has subsequently been more fully developed. Research in the early 1990s showed that an eye like ours could evolve in a series of small stages (in theory, they could be infinitely small stages) from an initial light-sensitive cell. In fact, therefore, the eye poses no problem with a need for several coordinated changes at the same time. Its evolution can proceed step-wise, with each change in a single part being advantageous by itself.

The eye is not the only organ that can be argued about in this way. In general, Darwin concludes, 'We should be extremely cautious in concluding that an organ could not have been formed by transitional gradations of some kind.' For most organs, there is comparative evidence suggesting the simpler beginnings of the organ. But even if we do not know of comparative evidence, and cannot think of a simpler initial stage, that does not mean no such simpler stage existed. It may be our knowledge that is limited. Research has consistently shown that, if the matter is looked into, we can find a series of transitional stages leading to any complex organ. Darwin's basic argument about 'organs of extreme perfection' still holds up.

Modern research on eye evolution also illustrates another of Darwin's points. When we – or when many of us – first think about how an eye could evolve, it seems to require such a huge number of changes that (as each one occurred by a small random mutation that then had to spread by natural selection) it must have taken an almost impossibly long time. But it turns out that our imagination is a poor guide. The 1990s study I mentioned above showed that the whole process of eye evolution could occur in about half a million generations. In evolutionary terms, that is a very short time. Evolution on Earth has been going on for about 4,000 million years, and many life forms have generation times of less than a year. In the quoted passage Darwin says that the problem of eye evolution, 'though insuperable by our imagination, can hardly be considered real'. Elsewhere he says that problems of this kind are problems of the imagination, not of reason.

Part of the imaginative problem lies in the amount of time that evolution has been going on for. This is far longer than anything else in human experience. Another part of the problem is that we tend to underestimate the power of natural selection. Natural selection is powerful because it is cumulative. During the evolution of a complex organ, once the first stage is established it becomes the starting point for further improvements. For instance, once a pit of invaginated cells has evolved it may be advantageous to fill the pit with fluid. Later, some of the fluid may be replaced by a lens. Thus, the lens does not evolve by itself. It does not evolve while there is a flat surface of light-sensitive cells. It only starts to evolve after most of the eye structure is complete. The chance of evolving something as complex as an eye in one step is negligibly small. But if the evolution occurs in gradual stages, with each new stage adding to what has already been achieved, the probabilities become much easier. As we read Darwin's discussion of organs of extreme perfection, we find out about the power of piecemeal engineering.

4

HYBRIDISM AND BIODIVERSITY

The Origin of Species 4

The view generally entertained by naturalists is that species, when intercrossed, have been specially endowed with the quality of sterility, in order to prevent the confusion of all organic forms. This view certainly seems at first probable, for species within the same country could hardly have kept distinct had they been capable of crossing freely. The importance of the fact that hybrids are very generally sterile has, I think, been much underrated by some late writers. On the theory of natural selection the case is especially important, inasmuch as the sterility of hybrids could not possibly be of any advantage to them, and therefore could not have been acquired by the continued preservation of successive profitable degrees of sterility. I hope, however, to be able to show that sterility is not a specially acquired or endowed quality, but is incidental on other acquired differences. [. . .]

Considering the several rules now given, which govern the fertility of first crosses and of hybrids, we see that when forms, which must be considered as good and distinct species, are united, their fertility graduates from zero to perfect fertility, or even to fertility under certain conditions in excess. That their fertility, besides being eminently susceptible to favourable and unfavourable conditions, is innately variable. That it is by no means always the same in degree in the first cross and in the hybrids produced from this cross. That the fertility of hybrids is not related to the degree in which they resemble in external appearance either parent. [. . .]

Now do these complex and singular rules indicate that species have been endowed with sterility simply to prevent their becoming confounded in nature? I think not. For why should the sterility be so extremely different in degree, when various species are crossed, all of which we must suppose it would be equally important to keep from blending together? Why should the degree of sterility be innately variable in the individuals of the same species? Why should some species cross with facility, and yet produce very sterile hybrids; and other species cross with extreme difficulty, and yet produce fairly fertile hybrids? Why should there often be so great a difference in the result of a reciprocal cross between the same two species? Why, it may even be asked, has the production of hybrids been permitted? To grant to species the special power of producing hybrids, and then to stop their further propagation by different degrees of sterility, not strictly related to the facility of the first union between their parents, seems to be a strange arrangement. [. . .]

It may be urged, as a most forcible argument, that there must be some essential distinction between species and varieties, and that there must be some error in all the foregoing remarks, inasmuch as varieties, however much they may differ from each other in external appearance, cross

with perfect facility, and yield perfectly fertile offspring.
[. . .]

. . . the perfect fertility of so many domestic varieties,
differing widely from each other in appearance, for
instance of the pigeon or of the cabbage, is a remarkable
fact; more especially when we reflect how many species
there are, which, though resembling each other most
closely, are utterly sterile when intercrossed.

Darwin dedicated a chapter in *The Origin of Species* to
'hybridism'. Hybridism refers to what happens when two different species are crossed, or hybridized. Often the two species
fail to produce hybrid offspring, or hybrid offspring are produced but are themselves sterile. Modern evolutionary
biologists are mainly interested in the chapter because of its
relevance to the origin of the species. Many (if not all)
modern biologists define a species in terms of interbreeding:
a species is a group of organisms that are capable of interbreeding, and do not interbreed with members of other
species. On this concept of species, humans are one species
(formally, *Homo sapiens*) and chimpanzees (*Pan troglodytes*)
another, because humans breed with humans, and chimps
with chimps, but humans do not breed with chimps. The loss
of the ability to interbreed with another group of organisms is
the key event in the evolution of a new species. Sometime in
the past there was a single ancestral species. Then somehow,
over time, some members of that species evolve different
reproductive attributes from the other members of that
species. In this way, one species evolves into two.

However, for Darwin the topic of hybridism – that is, the
formation of hybrids by interbreeding between two species, or
two distinct varieties – was not so closely related to the origin

of new species. He was well aware of the importance of inter-breeding and non–interbreeding for the existence of species in nature. Indeed the second sentence of the quoted passage says that species would hardly exist, as distinct forms, if they were not prevented from interbreeding. Species had been defined in terms of interbreeding since well before Darwin's time. The British naturalist John Ray, for example, explicitly defined species as interbreeding units, and he was writing in the late 1600s. Darwin knew about this tradition, but he does not seem to have thought about species purely as interbreeding units. Humans differ from chimps not only in the absence of interbreeding, but also in their physical appearance. Humans and chimps look different. Darwin probably had a more flex-ible, broader concept of what a species is than do modern biologists who concentrate only on interbreeding. In Chapter 2, we saw how Darwin discussed the divergence of species: competition between more similar forms drives them apart over time. For Darwin, the evolution of differing forms was as much part of the origin of species as the loss of interbreeding. 'Hybridism' is the defining issue in the modern concept of speciation; but for Darwin it was only part of the process.

We can therefore read the passage on hybridism both to see what Darwin himself was concerned with, and what a modern biologist can mine from it. The two 'readings' need not be completely separate: Darwin was partially aware of our modern conception of species, and Darwin's own con-cerns are still interesting. However, it will be convenient to take the two readings in turn.

What, first, is Darwin's own argument? Darwin's first para-graph ends by saying that he aims to show that sterility between species 'is not a specially acquired or endowed qual-ity, but is incidental on other acquired differences'. The word 'acquired' can be read as 'evolved', and particularly as 'evolved by natural selection'. (As I've already said, Darwin did not use the words 'evolve' or 'evolution' in *The Origin of Species*, with one exception. He used other terms instead and here he uses 'acquire'.) Thus, the sterility of hybrids did not evolve by natural selection. The reason is that natural selection favours attributes that increase the survival and reproduction of organisms. Sterility by definition does the opposite – it prevents reproduction. Natural selection will act to reduce, or eliminate, sterility; it will not act to create it. This makes hybrid sterility, as he says, 'especially important' for his theory. Darwin was putting forward a theory of life, a theory in which the various features of living creatures have evolved by natural selection. And yet here we have a feature of life – a feature that is crucial for the existence of species – that cannot have evolved by natural selection. In Darwinian terms, sterility is not an adaptation. Two of Darwin's aims in the chapter are to show that sterility does not have the attributes of a true adaptation, and to explain how it can nevertheless evolve despite its adaptive disadvantage. The first of these two also take care of Darwin's other aim – to show that sterility is not a 'specially endowed' quality in hybrids. Here Darwin is referring to what we should now call creationism.

When Darwin was writing, most naturalists thought that each species was somehow separately created. Whatever the mechanism was, of this creation, it made sense to equip hybrids with sterility; otherwise the species would soon cease to exist. Sterility is then an adaptation, benefiting the species in much the same way as do standard examples of adaptations such as eyes, or wings. (We can ignore the questions of whether adaptations benefit species or individual organisms, because it was not a question these naturalists were concerned with. It mattered for Darwin, though, as we have seen and shall see again in this book.) Darwin needed to show that hybrid sterility was not an adaptation, both to save his own theory and to confound the creationists.

A classic example of an adaptation is something like an eye: it is a complex organ, the structure of which appears to be designed to form visual images. In practice, it is difficult to give an abstract definition of which features of living things are adaptations and which are not. Biologists to this day have not agreed on a definition of adaptation. Thus, when Darwin comes to argue that hybrid sterility is not an adaptation he cannot begin with some abstract criteria of non-adaptiveness and then work through the details. Instead he describes a series of properties of hybrid sterility that would be puzzling if it were an adaptation to keep species separate. The first property is variation. The form of hybrid sterility differs from one pair of species to the next, and varies between individuals within a species. To see the argument, contrast this with something like an eye. The form of the eye is pretty much the same across all vertebrate animals (fish, frogs, lizards, birds, mammals). This is because the laws of optical physics dictate that some structures work as eyes and others do not. The form of the eye does vary, if we look at animal life as a whole; but all the forms are clearly specialized structures designed for vision. They are not any old arrangement of cells. And the structure is strikingly constant across a wide variety of life, such as all the vertebrates. With hybrid sterility, by contrast, we find a range even within closely related forms. One pair of species might produce fully sterile hybrids, while another closely related pair of species produces fully fertile hybrids. All the degrees in between are found too. It is as if we found 'eyes' with the lens and retina one way round in humans, but the other way round in chimpanzees: it would make you wonder whether we had really been 'specially endowed' with these 'eyes' for seeing.

Extreme variability remains one of the criteria that biologists use to recognize non-adaptive attributes in living things. At a molecular level, many stretches of DNA appear to have no adaptive purpose; these regions of non-adaptive DNA are also often highly variable between individual organisms. The criterion is not certain, however; something could be variable and yet still adaptive. Darwin's argument is more suggestive than conclusive. However, he goes on to look at other properties of hybrid sterility, each of which would be 'a strange arrangement' if it were an adaptation. It all adds up to a strong case – and one that is widely accepted today. Practically no modern biologists think that sterility in hybrids has evolved to keep the parental species distinct.

Instead, Darwin argued, sterility evolved as a by-product, 'incidental on other acquired differences'. Darwin does spell this argument out, but not in the quoted passage. In modern terms, as two species evolve apart over time, they accumulate all sorts of differences. The new genes that have evolved in humans over the past five million years will differ from the new genes that have evolved in chimpanzees. It is not scientifically known what would happen if humans and chimps tried to hybridize, but it is likely that some of the newly evolved human attributes would prove incompatible with newly evolved chimp attributes when they were put together in a hybrid body. The resulting hybrid would be dysfunctional. The problem is much the same as if you mixed components between two makes of motor car. The resulting hybrid motor car would probably not work, because of incompatibilities between components. The failure of the hybrid car is not a design feature: engineers have not built incompatibility in to prevent hybridization and keep their make of car distinct. The hybrid failure is an incidental consequence; it just happens when two teams of engineers work independently. Likewise, when two species evolve independently for a while, they are likely to evolve to be incompatible. The degree of incompatibility will vary between species in whimsical ways. Some pairs of species will just turn out, as if by chance, to be compatible. Their hybrids will be fertile. Others will be deadly incompatible. The result depends on which 'components', or genes, have changed during evolution.

The way Darwin set the problem up, and his conclusion, correspond to modern thinking; but one complication has been added. Biologists now refer to the failure of two species to interbreed as 'reproductive isolation'. Two main kinds of reproductive isolation can be distinguished. One is the kind discussed by Darwin: the two species interbreed but the hybrids fail to breed. Alternatively, the two species may not interbreed to begin with. For instance, two species may have different courtship signals and fail to recognize each other as potential mates. Darwin probably ignored this second kind of isolation because the evidence available to him mainly came from the artificial cross-pollination of plants. No one did research on the events leading up to mating until the late nineteenth century – after Darwin had written the *Origin*.

Biologists still agree with Darwin that hybrid sterility evolves as an incidental by-product of evolutionary change in the parts of the organism. But there are two schools of thought about the prior kind of isolation, from such factors as courtship. Some biologists think it can evolve as a 'specially acquired quality' – that is, different courtship signals may evolve in the species, in order to avoid producing hybrids. Others think it, like hybrid sterility, evolves as a by-product. Courtship signals may evolve apart in two species for some reason other than the prevention of hybridization between the two species. When the courtship behaviour of the two species is sufficiently different, members of the two species will cease recognizing each other as potential mates. Thus, when modern biologists read Darwin's chapter, they see the same distinction (specially acquired quality or incidental on other acquired differences) that they are still concerned with. They then either agree in part, or completely, with Darwin's conclusion, depending on their view about reproductive isolation by courtship (and some related factors) – which Darwin did not consider.

Darwin next moves on to a further difficulty that hybrid sterility seemed to pose for his theory. It appeared, on a simple survey of the facts, that crosses between different species never produced fertile hybrids but that crosses between varieties within a species were always fertile. This is superficially damaging for an important claim in Darwin's theory, and also for his favourite argumentative device. In Darwin's theory, there is no big distinction between varieties and species. Varieties differ by a certain amount, and species differ by a larger amount. In Darwin's words, varieties are incipient species. The same process that gives rise to new varieties will, if continued for longer, give rise to new species. Darwin's theory therefore predicts that varieties will blur into species; there should be no clear criterion by which we can distinguish varieties from species.

And yet it seems that inter-fertility might provide such a criterion. Varieties, it seems, might be different forms that are inter-fertile, and species different forms that are not inter-fertile.

Varieties might then be different kinds of entities from species, and it would be unsafe to reason from the one to the other. However, that is exactly how Darwin did like to reason. He repeatedly made an analogy between the kind of artificial selection that humans practise, when producing new domestic and agricultural crops and breeds, and the kind of selection that must occur in nature. But the domestic varieties produced by human action remain just that – varieties. Darwin notes (in the quoted passage) that it is 'a remarkable fact'. Pigeon varieties that would be classified as separate species on the basis of their appearance are nevertheless inter-fertile. A critic who thought that species were separately created might seize on this fact. Maybe the kind of evolutionary processes discussed by Darwin could operate on a small scale, within a species; but they do not seem to go far enough to produce new reproductively distinct species.

Darwin raised several objections. One is that it is largely a matter of definition, that we call things varieties if they are inter-fertile but species if they are not. The underlying reality may show a less clear distinction. Darwin was able to point to some evidence, in which 'varieties' within a species showed some reduced inter-fertility and in which 'species' were partly inter-fertile. However, the evidence was limited in his time, and it did not fit neatly with his theory that so many domestic varieties – of crops, of dogs, of pigeons – showed no sign of reduced fertility in crosses between extremely different forms. His theory clearly predicts that reduced fertility should evolve, at some point, between varieties that have evolved apart.

Modern evidence supports Darwin's prediction better than the evidence he had at the time. Biologists have now conducted many experiments in which they select different subsamples from an initial population in different ways, such that the subsamples evolve apart over time. Then, after a number of generations, we can measure whether any reproductive isolation has evolved between them. In fact a degree of reproductive isolation usually does evolve; but it takes a carefully designed experiment to detect it. The observations that Darwin drew on were probably too crude to be able to reveal the subtle reductions in inter-fertility between varieties. It remains an important Darwinian claim that varieties are not qualitatively distinct from species. This passage in the *Origin* is therefore important for the way it set the problem up. It also identifies the right kind of evidence to use to test the claim. However, Darwin seems to have run into difficulties because the evidence that was available to him was too crude and limited.

5

THE GEOLOGICAL SUCCESSION

The Origin of Species 5

. . . the imperfection in the geological record mainly results from another and more important cause than any of the foregoing: namely, from the several formations being separated from each other by wide intervals of time. When we see the formations tabulated in written works, or when we follow them in nature, it is difficult to avoid believing that they are closely consecutive. But we know, for instance, from Sir R. Murchison's great work on Russia, what wide gaps there are in that country between the superimposed formations; so it is in North America, and in many other parts of the world. The most skilful geologist, if his attention had been exclusively confined to these large territories, would never have suspected that during the periods which were blank and barren in his own country, great piles of sediment, charged with new and peculiar forms of life, had elsewhere been accumulated. [. . .]

On the sudden appearance of groups of Allied Species in the lowest known fossiliferous strata. There is another and allied difficulty, which is much graver. I allude to the manner in which numbers of species of the same group suddenly appear in the lowest known fossiliferous rocks. Most of the arguments which have convinced me that all the existing species of the same group have descended from one progenitor, apply with nearly equal force to the earliest known species. For instance, I cannot doubt that all the Silurian trilobites have descended from some one crustacean, which must have lived long before the Silurian age, and which probably differed greatly from any known animal. [. . .] Consequently, if my theory is true, it is indisputable that before the lowest Silurian stratum was deposited, long periods elapsed, as long as, or probably far longer than, the whole interval from the Silurian age to the present day; and that during these vast, yet quite unknown, periods of time, the world swarmed with living creatures.

On Extinction. We have as yet spoken only incidentally of the disappearance of species and of groups of species. On the theory of natural selection the extinction of old forms and the production of new and improved forms are intimately connected together. The old notion of all the inhabitants of the earth having been swept away at successive periods by catastrophes, is very generally given up, even by those geologists, as Elie de Beaumont, Murchison, Barrande, &c., whose general views would naturally lead them to this conclusion. On the contrary, we have every reason to believe, from the study of the tertiary formations, that species and groups of species gradually disappear, one after another, first from one spot, then from another, and finally from the world. [. . .] With respect to the apparently sudden extermination of whole families or orders, as of Trilobites at the close of the palaeozoic period and of

Ammonites at the close of the secondary period, we must
remember what has already been said on the probable wide
intervals of time between our consecutive formations; and in
these intervals there may have been much slow extermination.

The period when Darwin was developing his theory of
evolution was also the climax of one of the greatest scientific
research programmes: the mapping out of geological history.
Geological history is divided into a series of named periods –
from the present back to about 65 million years ago is the
Cenozoic (or Cainozoic), from about 65 back to 250 million
years ago is the Mesozoic, and from about 250–540 million
years ago is the Palaeozoic. These three periods cover the
main fossil record. Before the Palaeozoic is a long period
reaching back to the origin of the Earth about 4,500 million
years ago; it is subdivided in various ways but is often simply
referred to as the Precambrian. The three large periods
(Palaeozoic, Mesozoic, Cenozoic) are in turn subdivided: the
Palaeozoic is subdivided into Cambrian, Ordovician, Silurian,
Devonian, Carboniferous and Permian, for example. Between
the mid-eighteenth and mid-nineteenth centuries, geologists
learned to recognize the characteristic rocks of these geological periods. Mesozoic rocks, for instance, contain invertebrate
animals such as the now-extinct group of ammonites, as well
as vertebrate animals such as fish and reptiles, but very few
mammals. Geologists also worked out the historical order of
the periods and, as they did so, they described the rise and fall
of various fossil groups during the history of life.

The research concentrated on what were called 'formations' – a certain rock type (such as sandstone or limestone) in
a certain locality. In special places such as cliffs, a series of distinct formations could be recognized, one above the other;

but in most places only one formation is visible – the rock type below the topsoil. The reconstruction of history requires tracing how one characteristic formation graded into another; but this could not be done in any one region because not all the formations would be preserved there. The key method was called 'correlation': geologists looked for equivalent formations in different countries, by comparing the rocks and the fossil composition; this comparison was called 'correlating' the two formations. One region might then seem to have three successive formations, which we can call A, C and E. Another region might also have formations A and C, but with another formation, B, between them. We thus build up the full sequence as A, B, C, E. As more places are studied, a more comprehensive and reliable picture of history is built up.

During the late eighteenth and early nineteenth centuries, the more recent formations, from the Carboniferous to the present, had been worked out. Research then moved on to the more difficult, older rocks. During the 1830s and 1840s the modern system for what is now called the Palaeozoic was being formulated. The main centre for reporting and discussing the work was the Geological Society of London, of which Darwin was a very active member after his return from the *Beagle* voyage in 1836; indeed it was practically the centre of his social life at that time. After his marriage in 1842 he continued to be an active member.

The facts of geological history could be used to test the theory of evolution in various ways, as we shall see in this chapter and the next. But the quoted passage begins with a discovery that had emerged when formations were correlated from place to place: the rocks at any one place contained a very incomplete record of geological history. This was not just an assertion, or deduction from theory; it was a common –

indeed universal – result in the kind of research Darwin was
thinking about every day. When formations are correlated
between two places, they are never exactly the same: one formation might have a deeper expanse of rock in one place
than the other, and a whole formation that is present in one
place may be absent from the other. Sir R. Murchison
(Roderick Impey Murchison), mentioned near the beginning of the quoted passage, was a leader in research on
geological history, and Darwin knew him personally.
Murchison had travelled to Russia in search of rocks between
the Carboniferous and the Triassic – two periods well repre-
sented in British rocks. Near Perm in the Urals, he found
formations that fitted in between; these formations are now
called Permian. There are a few rocks of Permian age in
Britain, but they are too few for the period to be worked out
from research in Britain alone. Murchison's knowledge of
British geology had in turn allowed him to recognize that
there were gaps between the rock formations he found in
Russia.

The gaps in the geological record were significant for two
features of history that Darwin goes on to discuss in the
quoted passage: the sudden appearance, and the sudden
extinction, of groups of species. One word may be puzzling in
Darwin's discussion of the first problem: the word 'Silurian'.
In modern geology, the Silurian period dates from 443–417
million years ago, almost halfway through the Palaeozoic.
Many scientific readers would expect Darwin to have said
'Cambrian' instead. The Cambrian is the earliest period
within the Palaeozoic, dating from 542–490 million years ago.
In Darwin's time, and for about another century after he
wrote, no fossils were known from rocks older than the
Cambrian: the Cambrian contained (in Darwin's terms) 'the

lowest fossiliferous strata'. It is still the case that the great
majority of known fossils date from times from the Cambrian
to the present.

So why did Darwin say Silurian? Part of the answer is that
the term Silurian, as introduced by Murchison, includes the
Silurian both in the modern sense and the previous period,
the Ordovician. The other part of the answer is that the term
Cambrian was controversial. Geologists had not yet learned to
recognize Cambrian rocks, as distinct from Silurian rocks,
and Darwin was probably steering clear of a predictable row as
he avoided the term 'Cambrian', though that is the term we
should use now.

The evidence in Darwin's time suggested that several distinct groups of animals all arose relatively suddenly in the
earliest fossil-containing formations. With some qualification,
the evidence still suggests something similar today. The phe-
nomenon is called the Cambrian explosion. For Darwin, the
sudden appearance of an animal group posed a problem. As he
said in a nearby section, 'If numerous species, belonging to the
same genera or families, have really started life all at once, the
fact would be fatal to the theory of descent with slow modification through natural selection.'

His solution is the factor that the quoted passage begins
with: missing formations. In particular, he is led to postulate
(or to predict) a huge, unknown stretch of time before the
earliest trilobites, during which slow evolution could have
led from the simplest initial life forms to the early fossils (such
as trilobites) in the Cambrian – or Silurian in Darwin's sense.
(Trilobites are a fossil group of animals; they are more closely
related to spiders than to crustaceans such as crabs and
shrimps. However, spiders, trilobites and crustaceans are all
related – they are all arthropods – and the still unknown

ancestor of the trilobites could have resembled some crustacean form.)

In fact Darwin was right: if we take his 'lowest fossiliferous stratum' to refer to the base of the Cambrian, about 540 million years ago, then there is indeed a far longer period before it when the world at least contained (if it may not have 'swarmed with') living creatures. The earliest evidence of life from around 3,500–3,800 million years ago. The Precambrian fossil record is therefore about six times as long as the period from the Cambrian to the present.

Although the Precambrian fossil record has now been discovered, something of a puzzle remains. That record is remarkably thin; the first Precambrian fossil discoveries were not made until about the year 1950, and since then there has been a steady trickle of further finds. Fairly large fossil creatures have been found just before the base of the Cambrian, at about 560 or even 580 million years ago in Australia and in China, but they do not tell us much about the prehistory of Cambrian forms of life such as trilobites. It would be at least convenient for Darwin's theory if the Precambrian fossils revealed some of the ancestral stages that preceded the Cambrian fossils. But Precambrian fossils are mainly simple, single-celled life forms, or if they are more complex then they do not seem to be closely related to later animals; they do not blur away the impression that the early animals appear suddenly. One currently popular idea (though it is only an idea, not a firm conclusion) is that the Cambrian was the time when 'hard parts' originated. Hard parts are things like bones in vertebrated animals (such as ourselves), shells in molluscs, and the hard carapaces of animals like crabs and trilobites. Hard parts are much more likely to be preserved as fossils than are the soft parts of living things. Then, the sudden appearance of fossil animals about 540 million years ago means not that animals evolved suddenly – Darwin was right about the long phase of prior evolution before the Cambrian – but that something at that time made the evolution of hard parts advantageous. And there are, in turn, several hypotheses about what that 'something' was. One popular hypothesis is that predation may have become more powerful in the late Precambrian, and hard parts were needed for defence. In earlier times, the main 'predators' may simply have grazed on smaller creatures, rather in the way that whales now filter smaller creatures out of the water. Hard parts are not much of a defence against that fate. But in the late Precambrian and early Cambrian, predators with powerful claws and jaws may have arisen. External armour can help to fend off predators of this kind.

Darwin's second problem was sudden extinction. Darwin had developed his geological views under the influence of Charles Lyell's theoretical system. Lyell's opponents called him a 'uniformitarian', and the word has stuck. Uniformitarians explain the geological past in terms of processes that observably operate in the world today; they do not invoke hypothetical, unobservable factors. When Darwin was writing, a 'catastrophist' school of thought was in something of a decline, as Lyell's uniformitarianism spread its influence. Catastrophists claimed that, during the history of life, there was a series of catastrophes, in which all the species on Earth went extinct and then a new set of species was created. In popular thinking, the most recent of these catastrophes was more or less identified with the biblical Flood.

Darwin begins by noting that the 'old notion' of catastrophic extinction had been 'very generally given up'. He then says that we have good evidence for gradual, rather than catastrophic, extinctions. Darwin reasoned that, on the theory of natural selection, extinctions should be gradual. The quoted passage gives only a hint of his idea, but he spells it out in more detail in a nearby passage. He argues that extinction occurs when a new species arises that has 'some advantage over the [other species] with which it comes into competition'. The inferior form will gradually decline as the new species spreads through its geographic range. Such was the process that Darwin had in mind when he says (in the quoted passage) that the extinction of old forms is intimately connected with the production of new forms.

If Darwin's theory of extinction is correct, it is puzzling that whole groups of species seem to become extinct suddenly and simultaneously. The fossil record contains examples, as Darwin admits: for instance, trilobites at the close of the Palaeozoic and ammonites at the end of the secondary. Again, Darwin's chronological terms are not quite the same as those now in use. The Palaeozoic is the same as now; it is the first of three big phases – the Palaeozoic, Mesozoic and Cenozoic. But another, older set of terms overlapped with these: the Primary, Secondary and Tertiary. (Sharp-eyed readers will notice that the conventions about capital letters have changed between Darwin's time and ours.) Simplifying somewhat, the Primary more or less corresponded to the Precambrian and the Secondary to the time from Murchison's Silurian to the end of the Cretaceous. The Tertiary is still used; it is a major chunk of the Cenozoic. Thus, the extinctions at the end of the Secondary were in modern terms at the Cretaceous-Tertiary (or K/T) boundary; this was the time when the dinosaurs went extinct.

Darwin's first explanation for these 'apparently sudden exterminations' is gaps in the fossil record. There could be a missing rock formation during which there was a gradual decline of the ammonites, or of the trilobites. This was a plausible explanation at the time, but it is less so now, because we have absolute dates for many rocks. Radioisotopes can be used to establish the dates of rocks from the end of the Cretaceous and the beginning of the Tertiary. What we find is that the rocks, at least at some sites, show a continuous sequence of times; there is no hiatus between the Cretaceous rocks and the Tertiary rocks. The extinctions at the time really do seem to have been sudden. Since 1980, the sudden extinctions have been associated with a collision between the Earth and an extraterrestrial asteroid.

Most modern biologists and geologists hold different views about extinction from Darwin. He was sceptical about the reality of mass extinctions; he explained extinctions in terms of competition between new, superior species and old, inferior species. Most modern scientists accept that at least some mass extinctions have occurred. They also accept that some extinctions have been caused by biological competition. However, opinions vary about both of these factors. The number of mass extinctions that have occurred during the history of life is uncertain. Estimates range from as few as 2 to as many as 13. Darwin's basic explanation – fluctuations in the sedimentary record – may explain some of the controversial apparent mass extinctions, even if it does not explain all of them. The importance of biological competition in causing extinction also remains unknown, not least because it is so hard to study in fossils.

Although modern scientists think that 'sudden exterminations' are more important in extinction, and natural selection less, than Darwin did, I doubt whether he would have disapproved of the change. Nothing deep in his theory has been challenged. Darwin, like Lyell, objected to catastrophic explanations because they invoked processes that could not be scientifically studied. In modern science, several lines of objective evidence can be used to infer catastrophic extraterrestrial impacts and calculate the consequences. Darwin would probably, like most modern thinkers, have seen this as a welcome addition to his basic theory.

6

THE CASE FOR EVOLUTION

The Origin of Species 6

If we admit that the geological record is imperfect in an extreme degree, then such facts as the record gives, support the theory of descent and modification. [. . .] The fact of the fossil remains of each formation being in some degree intermediate in character between the fossils in the formations above and below, is simply explained by their intermediate position in the chain of descent. The grand fact that all extinct organic beings belong to the same system with recent beings, falling either into the same or into intermediate groups, follows from the living and the extinct being the offspring of common parents. [. . .]

Looking to geographic distribution [. . .] We see the full meaning of the wonderful fact, which must have struck every traveller, namely, that on the same continent, under the most diverse conditions, under heat and cold, on mountains and lowland, on deserts and marshes, most of the inhabitants within each great class are plainly related; for they will generally be descendants of the same progenitors and early colonists. [. . .]

The fact, as we have seen, that all past and present

organic beings constitute one grand natural system, with group subordinate to group, and with extinct groups often falling in between recent groups, is intelligible on the theory of natural selection with its contingencies of extinction and divergence of character. [. . .]

The framework of bones being the same in the hand of a man, wing of a bat, fin of the porpoise, and leg of the horse, – the same number of vertebrae forming the neck of the giraffe and of the elephant, – and innumerable other such facts, at once explain themselves on the theory of descent with slow and slight successive modifications. The similarity of pattern in the wing, and leg of a bat, though used for such different purposes, – in the jaws and legs of a crab, – in the petals, stamens, and pistils of a flower, is likewise intelligible on the view of the gradual modification of parts or organs, which were alike in the early progenitor of each class. On the principle of successive variations not always supervening at an early age, and being inherited at a corresponding not early period of life, we can clearly see why the embryos of mammals, birds, reptiles, and fishes should be so closely alike, and should be so unlike the adult forms. We may cease marvelling at the embryo of an air-breathing mammal or bird having branchial slits and arteries running in loops, like those in a fish which has to breathe the air dissolved in water, by the aid of well-developed branchiae.

Disuse, aided sometimes by natural selection, will often tend to reduce an organ, when it has become useless by changed habits or under changed conditions of life; and we can clearly understand on this view the meaning of rudimentary organs.

Darwin's second main purpose in *The Origin of Species* was to argue for evolution. Unlike the argument for natural selection, Darwin here had a clear alternative to argue against – creationism, the idea that each species has a separate origin and remains constant in form after its origin. In Darwin's time, as in ours, the main inspiration for creationism was religious. Most creationists think that each species was separately created by God and has remained unchanged ever since. Darwin, however, tends to leave God out. He treats creationism as a scientific hypothesis, which suggests that each species has an independent origin. This possibility does not depend on whether those origins are due to natural or supernatural mechanisms. In scientific discussion, God tends to be a fairly useless hypothesis: simply saying 'God did it' adds little to the discussion. The few passages where Darwin raised the question make it clear that he thought it empty to invoke God (or some euphemism for God), in a scientific argument.

Evolution differs from creationism in two respects. One is that, according to the theory of evolution, species change over time. The ancestor of a modern species will differ from the modern form, if we trace it far enough back. The other is that, according to the theory of evolution, modern species have descended from shared ancestors in the past. Darwin's argument for evolution concentrates on this second point. In this respect his argument differs from many modern presentations of the case for evolution. Modern biologists usually point to examples of observable evolutionary change in the short term, together with the kind of similarities seen between species that Darwin discussed and that suggest common ancestry. For instance, we can see the evolution of drug resistance in HIV (the virus that causes AIDS) within 2 to 3 days in an AIDS patient. Darwin did not have evidence of this kind.

Evidence of evolution in action was not discovered until about the 1920s, and has been accumulating since then. Evolution is usually too slow to observe, but in exceptional cases (such as a population of viruses that are being blasted with drugs designed to kill them) it can be strikingly fast. Moreover, we now have examples of evolutionary change in a series of fossil populations over time. The examples are rare, because it is unusual to find a series of populations preserved in the fossil record; but they do exist and are clear evidence of evolution. Darwin would no doubt have been delighted to have these two kinds of evidence, of change in living and in fossil populations, but in their absence he concentrated on evidence of shared ancestry.

Darwin makes his case for evolution in a series of chapters in the *Origin*: two chapters on the fossil evidence, two chapters on geographical distribution, and so on. The whole argument is admirably clear and remains the most interesting account of the case for evolution. Other authors have made much the same argument, but they lack Darwin's master intellect and historical force. After the main chapters, Darwin finished with a summary chapter, which provides a quick overview of the whole book. I have extracted the quoted passages from that summary chapter.

Although Darwin's argument is, in modern terms, between evolution and creationism, he uses neither word. As I said in Chapter 1, Darwin's term for evolution is 'descent with modification'. He does use the word 'evolved' once in *The Origin of Species* – it is the last word in the book – but Darwin and others only later came to use the word 'evolution' in the way we now use it. 'Creationism' is a modern word, though Darwin does use related expressions. One potential difficulty for a first-time reader is that he does not say them all that often. Creationism is implicitly the target of Darwin's argument in about half the chapters of the *Origin*. However, Darwin often only interprets each piece of evidence in terms of evolution; he does not always explain in detail how the same piece of evidence undermines creationism. He may leave it to the reader to think through how the evidence would look if species had separate origins. Creationism is particularly concealed in the final chapter that I have quoted from, where Darwin simply gives the positive evolutionary side of the argument. A full appreciation of his case requires us to think about the negative case that lies implicit in what he writes.

Darwin begins with the fossil record. The simplest fossil evidence for evolution would probably be a series of forms, in which one transformed into the other over time. However, the fossil record is usually too incomplete to provide this. Darwin instead drew attention to several other features of the fossil record that only make sense if the theory of evolution holds good. I have quoted two examples. The first is the way intermediate forms tend to occur at intermediate times in the fossil record. An example comes from our own ancestry, though this is not the example Darwin gives. Our ancestors, around 400–500 million years ago, were fish. At some point, amphibians (relatives of such modern creatures as frogs) evolved from a certain group of fish. Amphibians live partly in water, and partly on land. From amphibian ancestors, in turn, the reptiles evolved and lived exclusively on land. Mammals subsequently evolved from reptiles, and we are mammals. Evolution proceeded in the direction FISH → AMPHIBIANS → REPTILES → MAMMALS. If we look carefully at members of these four groups, we can see that amphibians are in many respects intermediate between fish and reptiles. Amphibians, for instance, sometimes breathe via gills, like fish, and sometimes via lungs. But they lack a rib cage for filling their lungs with air. Thus we can deduce, from looking at the anatomy of the modern forms, that if evolution is to proceed from a fish to a reptile stage, it almost has to go through something like an amphibian stage. This being so, it would be surprising (on the theory of evolution) if reptile fossils came first, and then were followed by fish fossils, with amphibians coming last. We can predict, from their anatomical structure, that they evolved in the order fish → amphibian → reptile. The actual order of the fossils is what we predict: the intermediate forms are found at intermediate times. The argument is not completely convincing with three groups, such as fish, amphibian and reptiles; but here I am only illustrating the logic of the argument. When we add a long series of animal groups, and their order in the fossil record matches their anatomical relationships, the argument becomes powerful.

By contrast, if fish, amphibians and reptiles had all been independently created, there is no reason to expect the anatomically intermediate group to appear at an intermediate time in the fossil record. A creationist can only explain the match by chance.

In Darwin's second argument, he is implicitly arguing against a version of creationism that had strong supporters in his time but is no longer popular. He remarks that extinct forms fall into the same classificatory groups as do living forms; in other words, extinct forms are not completely unrelated to modern life. As we saw in Chapter 5, geologists in the years before Darwin had suggested that the history of life consisted of a series of catastrophic extinctions, in which all life was wiped out, each followed by a round of creations in which new life forms arose. If this was correct, we should expect the extinct fossils from one of these earlier phases to be unrelated to modern life. Modern life could be traced back to the most recent round of creation, following the extinction of earlier life. In fact, extinct fossil forms are recognizably related to modern life. Creationism comes in many versions, and when arguing against creationism it is necessary to tune the evidence to the particular version your reader may support. Modern creationists do not advocate a series of rounds of creation, following catastrophic extinctions, although a school of creationists in Darwin's time advocated just that, and Darwin had to take them into account.

The second class of evidence comes from the geographical distributions of living things. (Again, Darwin looked at several kinds of evidence, but I have only quoted one of them as an example.) If we look at life in a particular region of the globe, we often find that the species living there are closely related – more closely related than would be expected if each species in an area had been separately created. The most famous Darwinian example is the group of birds now known as 'Darwin's finches', which live on the Galápagos Islands. There are about twelve species, all closely related. They are classified together as a distinct group of species, which means that each of Darwin's finches is more closely related to the other Darwin finches than to any other species of life on Earth. And yet the group of species contains a diversity of life forms: some, for instance, live like finches and pick up seeds as food; another has evolved as a sort of woodpecker. True wood-peckers have long probing beaks and tongues that they use to extract insects from tree bark. The Galápagos 'woodpecker' finch uses sticks that it probes into tree holes, to feed some-what like a woodpecker. There are no normal, or 'true', woodpeckers on the Galápagos. On the theory of evolution, the facts make sense. In the past, an ancestral finch colonized the Galápagos. There were no (or few) other birds on the islands. That ancestral species duly evolved into a number of descendant species, with a diversity of ways of life including a 'woodpecker' one.

However, if species are independently created, the facts make less sense. Elsewhere on Earth, woodpeckers fill the woodpecker way of life. If woodpeckers work elsewhere, why not create a woodpecker in the Galápagos? Why create a modified finch, that happens to be very like the other finches of the islands? Much the same argument can be made for all over the globe. It was particularly important for Darwin: evidence from geographical distributions (and not from fossils or the topics we are coming to below) first persuaded him about evolution.

Darwin's next argument concerns what he calls the 'natural system'. 'System' was used in the eighteenth and nineteenth centuries where we now say 'classification' or 'taxonomy'. (The word survives in 'systematics', which is the science of biological classification.) The classification of life is hierarchical, 'with group subordinate to group', as Darwin says. That is, subgroups such as cats or monkeys are contained within the larger group of mammals. This hierarchical structure is what we should expect if life has evolved from a common ancestor. Indeed the hierarchical classification in a sense matches the branching structure of life's history. If we trace back from modern cats, we soon come to a common ancestor of all cats. If we trace back from that ancestor, in time we come to a common ancestor of all mammals . . . and later the common ancestor of all animals, and finally of all life on Earth. But a hierarchy of groups within groups is exactly what we should *not* expect to find if each species is separately created. Each species will have its own individual attributes. If we classified them, the classification could take on almost any form, depending on the creation process. Maybe the species would fit a system like the Periodic Table in chemistry, or merely like the alphabetic index of a book. There is no particular reason to expect a hierarchical 'natural system' if each species is independently created.

Darwin then moves on to what most modern biologists would regard as the most powerful evidence of evolution – from homology. Indeed, all Darwin's classes of evidence are arguably different forms of a general argument from homology. Homology is a little difficult to define here, because it is now usually defined in terms of evolution. A homology is a characteristic present in two species that was also present in their common ancestor. Backbones, for example, are present in both human beings and chimps. The common ancestor of the two species also had a backbone. Backbones in humans and chimps are an example of a homology. Homologous similarity is ancestral similarity – similarity due to structures inherited from a common ancestor. However, the modern definition is an evolutionary interpretation of facts that were known in pre-evolutionary biology. Homology then referred to similarities between species, including similarities that could not easily be explained given the way of life in the species.

Darwin's example is the human hand and the homologous structure in other related species. The hand has five digits, and a certain arrangement of bones. We use our hands for manipulating and gripping objects. However, the same five digits and bone arrangement is found in a bat wing and a porpoise fin, and in modified form in a horse's foot – even though those structures are used in a very different way from the human hand. It seems unlikely that the same number of digits and the same bone arrangement is needed in all the species. This similarity of structure is homologous similarity. It suggests that evolution has happened, because if porpoises, bats, humans and horses were separately created, they would not all have the same basic limb structure. Given the differences in the way they use their limbs, they would have been created with different designs. Likewise, Darwin discusses similarities between the embryos of different species, such as the apparently fish-like stages in our own embryonic development. He also discusses similarities between the rudimentary organs of some species and the fully developed organs in other species.

Since Darwin's time, biologists have continued to find homologous similarities between life forms. The most striking examples come from molecular biology. The molecular DNA contains a set of coded instructions for building a body. The code in which the instructions is written is called the genetic code. The genetic code is arbitrary in much the same way that human language is arbitrary (that is, there is no particular reason why the sequence of letters H-U-M-A-N should refer to what in fact it does). And yet, it turns out, all life uses essentially the same genetic code. If life has evolved from a common ancestor, this makes sense. That common ancestor hit on a particular code and since then it has been handed down to all life. But if each species were independently created, there would be no reason for them all to use the same genetic code. It would be as surprising as if intelligent life forms, independently evolved throughout the universe, all turned out to speak English.

When Darwin wrote, he knew of homologies that linked a wide variety of life forms: for instance, homologous bones are shared between all fish, amphibians, reptiles, birds and mammals; all these creatures, therefore, probably shared a common ancestor. But Darwin lacked 'universal' homologies that were shared between all life forms. He would have been delighted by the findings of universal molecular homologies such as the genetic code. They are the best evidence we have that all life on Earth diversified from a common ancestor.

7

THE SOCIAL AND MORAL FACULTIES

The Descent of Man 1

The Descent of Man or, to give its full title, *The Descent of Man, and Selection in Relation to Sex* is Darwin's second most important book. It seems almost like two books accidentally bound together. Approximately a third of the book is about human evolution; it considers the evidence that humans have evolved from ape-like ancestors, and has chapters on the evolution of mental, moral and social faculties. The other two-thirds of the book is about what Darwin called 'sexual selection'. Sexual selection is Darwin's theory to explain sex differences in all life: why males tend to fight more than females do, and why males in some species have ornaments such as bright feathers. The part of the book on sexual selection works through a long list of non-human forms of life, and has nothing to say about human evolution at all. Only in a final, short section of the book does Darwin tie together his two themes. He argues that sexual selection may explain human racial difference, in skin colour and facial appearance. Some commentators regard the final, linking section as little more than a gesture of unity between two essentially separate books. Others, however, say we should take Darwin's design seriously and see the fundamental unity of the whole. I shall not attempt to solve the problem here. In this and the next two chapters, I have selected three extracts, two about human evolution, followed by one about sexual selection. The first two are closely related, but little related to the third.

When two tribes of primeval man, living in the same country, came into competition, if (other circumstances being equal) the one tribe included a great number of courageous, sympathetic and faithful members, who were always ready to warn each other of danger, to aid and defend each other, this tribe would succeed better and conquer the other. Let it be borne in mind how all-important, in the never-ceasing wars of savages, fidelity and courage must be. The advantage which disciplined soldiers have over undisciplined hordes follows chiefly from the confidence which each man feels in his comrades. [. . .] Selfish and contentious people will not cohere, and without coherence nothing can be effected. A tribe rich in the above qualities would spread and be victorious over other tribes: but in the course of time it would, judging from all past history, be in its turn overcome by some other tribe still more highly endowed. Thus the social and moral qualities would tend slowly to advance and be diffused through the world.

But it may be asked, how within the limits of the same tribe did a large number of members first become endowed with these social and moral qualities, and how was the standard of excellence raised? It is extremely doubtful whether the offspring of the more sympathetic and benevolent parents, or of those who were the most faithful to their comrades, would be reared in greater numbers than the children of selfish and treacherous parents belonging to the same tribe. He who was ready to sacrifice his life, as many a savage has been, rather than betray his comrades, would often leave no offspring to inherit his noble nature. The bravest men, who were always willing to come to the front in war, and who freely risked their lives for others, would on an average perish in larger numbers than other men. Therefore it hardly seems probable that the number of men gifted with such virtues, or that the standard of their excellence, could be increased through natural selection, that is, by the survival of the fittest; for we are not here speaking of one tribe being victorious over another.

Although the circumstances, leading to an increase in the number of those thus endowed within the same tribe, are too complex to be clearly followed out, we can trace some of the probable steps. In the first place, as the reasoning powers and foresight of the members became improved, each man would soon learn that if he aided his fellow-men, he would commonly receive aid in return. From this low motive he might acquire the habit of aiding his fellows. [. . .] Another and much more powerful stimulus to the development of the social virtues is afforded by the praise and blame of our fellow-men. [. . .] It is hardly possible to exaggerate the importance during rude times of the love of praise and dread of blame. A man who was not impelled by any deep, instinctive feeling, to sacrifice his life for the good of others, yet was roused to such actions by a sense of glory, would by his example excite the same wish for glory in other men and would strengthen by exercise the noble feeling of admiration. He might thus do far more good to his tribe than by begetting offspring with a tendency to inherit his own high standards. [. . .]

It must not be forgotten that although a high standard of

morality gives but a slight or no advantage to each individual man and his children over the other men of the same tribe, yet that an increase in the number of well-endowed men and an advancement in the standard of morality will certainly give an immense advantage to one tribe over another. A tribe including many members who, from possessing in a high degree the spirit of patriotism, fidelity, obedience, courage, and sympathy, were always ready to aid one another, and to sacrifice themselves for the common good, would be victorious over most other tribes; and this would be natural selection.

Darwin is here concerned with the evolution of what modern biologists call altruism, self-sacrificial behaviour in which one individual helps another. In more exact terms, altruism is behaviour that has a cost to the altruist but a benefit to the recipient. The section I have quoted from is a marvel for the modern theorist of altruism, because it identifies all of what we still think of as the basic problems and all but one of the currently recognized solutions.

The question is how humans could have evolved all the kinds of cooperative behaviour that we see in human societies, and on which society depends. Darwin's argument, that these 'social and moral faculties' have evolved by natural selection, was particularly controversial in his time. On a religious view, our mental ability and moral sense are what distinguish us from the brute creation: our bodies may in part resemble animals, but our moral sense is quite unlike anything seen in non-human animals and is a divine attribute, peculiar to humans among Earthly life forms. Darwin replied to this by tracing rudiments of morality in non-human animals, and by showing how morality could evolve by stages in human beings.

Darwin identifies the advantage of cooperation, and so of the moral sense that makes human cooperation possible, in war. If two tribes were competing, the tribe whose members cooperated better could be expected to win. A tribe of selfish individuals would soon be conquered and eliminated. Morality has advanced by the advantage it has given in battle.

This may be so, at least in part, but still there is a paradox. How can natural selection favour an individual who lays down his life for his tribe? The bravest individuals are more likely to be killed, and will for this reason produce fewer offspring on average. 'Therefore it hardly seems probable that the number of [such] men could be increased by natural selection.' To this day, discussions of altruism begin with this basic point. Any self-sacrificial behaviour seems in conflict with natural selection – which raises a question of how altruism could ever come to exist. This is the question Darwin is trying to answer in the remainder of the quoted passage. He suggests three possible answers.

The first is, or at least hints at, what would now be called reciprocity. If one man (*A*) helps another man (*B*) now, then *B* may help *A* later in return. Darwin differs from modern writers in basing reciprocity on rational calculation: on 'reasoning powers and foresight'. He calls it a 'low motive'. However, reciprocity does not require any reasoning powers or foresight at all. What it does require is individual recognition, or something that corresponds to it. For instance, the most famous modern study of reciprocal altruism looked at – vampire bats. Vampire bats live in roosts, containing several individuals. They fly out at night, in search of victims such as domestic animals, from which they take blood. On any one night, a particular bat may be unlucky and fail to find food. When it returns to the roost, a more successful bat may regurgitate a blood meal for the hungry individual. On a sub-sequent night, the roles may be reversed. A system of this kind requires several conditions: individual bats must be able to recognize each other, and to measure relative need. Otherwise the bat that gives away some of its meal may not be paid back later, except by chance, or it may direct its aid inappropriately. However, the system does not require any kind of rational cal-culation or foresight. Bats may be capable of both, but their mutual aid system of reciprocal blood regurgitation only needs bats to help one another after one bat has been successful and another has not. Natural selection favours the bat that gives away some of its meal, provided it is later paid back when it is in need. Darwin may have been aware of this general point (though not of the particular research on vampire bats). He says that, 'Each man would soon learn . . . that he would commonly receive aid in return.' This seems to me to imply some pre-existing reciprocity that can be learned about. Maybe Darwin was suggesting that reasoning and foresight could be used to develop further a system of reciprocity that already existed. However, a straightforward reading implies that Darwin based reciprocity on reasoning. Modern biolo-gists agree that reciprocity is one way in which natural selection can favour altruism, but would disagree that it has to be based on conscious reasoning.

Darwin then moves on to another factor, praise and blame from our fellow humans. He seems to me to be discussing how social or cultural factors, with or without natural selec-tion, could lead to self-sacrifice. I can see two main ways of spelling the argument out, in modern terms. In one, the self-sacrifice is not simply favoured by natural selection. Our sensitivity to praise and blame could have initially evolved by normal natural selection. Individuals who were sensitive to the expressed feelings of others in their society would produce more offspring. However, once that sensitivity had evolved, it could lead individuals to sacrifice themselves to attain greater social glory. (Alternatively, other members of society might tune their praise and blame in such a way as to manipulate an individual into self-sacrifice.) Then an individual may do something that natural selection works against, because of cultural influences. Another example would be religious celibacy. Celibacy − assuming it is associated with non-reproduction − is presumably opposed by natural selection. But individuals may still decide, because of their religious beliefs, to abstain from sex. Individual decisions, and cultural influences, may override natural selection.

Critics may object that, if our sensitivity to praise and blame, or our religious beliefs, lead to lowered reproduction then natural selection would have reconfigured our mental operations long ago. We could still be socially sensitive, or hold religious beliefs, but not in a way that contradicted natural selection. Natural selection will indeed have acted, and be acting, on our brain processes insofar as they are reproductively inefficient. However, Darwin's argument (on the interpretation we are exploring) can still work. Natural selection is a slow process relative to cultural change and individual human choice. Whatever mental processes we are equipped with, there will be ways for individuals to choose to do things that reduce their reproductive output. Natural selection is probably continually retuning our brains, but cultural factors do not stand still and they may also persistently lead some of us to behave in ways that reduce our reproduction.

There is a second way to interpret Darwin's argument, a way that denies there is any conflict between culture and natural selection. Maybe the individuals who seek the greatest social glory, by acts of self-sacrifice, on average gain from it. For every individual who dies, another individual who behaves in the same way may survive and reap his or her reward on Earth. If the gains of the survivor exceed the losses of the individual who dies, then on average it pays (in an evolutionary sense) to pursue glory in self-sacrifice.

These two interpretations are not the only ways in which culture, human decision-making, and natural selection may be related. Nor is it known which (if either) is correct – we are condemned to eternal uncertainty in our understanding of human behaviour. However, Darwin's argument can still stand (on more than one interpretation). Human decisions are influenced by cultural factors, and this may lead us to acts of self-sacrifice.

Finally, Darwin explains self-sacrifice in terms of what would now be called 'group selection'. A tribe whose members were more inclined to self-sacrifice would out-compete a tribe of more selfish, anarchic individuals. This advantage at the group (or tribe) level, says Darwin, 'would be natural selection'. An individual who sacrifices himself for the good of his tribe would lose out within the tribe, relative to the others who benefit from his sacrifice. But the advantage, at the level of the whole tribe, of having more members who sacrifice themselves, outweighs the individual disadvantage. Altruism evolves by natural selection because of its advantage to the group.

Most, but not all, modern evolutionary biologists allow group selection as a possible explanation for altruistic behaviour, but are sceptical whether the explanation works in reality. The reason is that when individual and group advantage are in conflict (as, for instance, in times of war), natural selection usually works more powerfully at the individual than the group level. Natural selection favours what is advantageous to the individual over the timescale of individual generations. When a selfish individual, who evades opportunities for self-sacrifice, produces more offspring than members of the tribe who sacrifice themselves for the common good, the frequency of selfish behaviour goes up. Traits that are advantageous to the tribe are favoured over the timescale of tribal 'generations'. That is, when a tribe of selfish individuals is killed off by a tribe of altruists, the frequency of altruism goes up. But the 'death' of a tribe is a rarer event than the death of an individual. Natural selection therefore usually ends up establishing whatever is favoured in the fast, persistent process of selection among individuals rather than what is favoured in the slow, intermittent process of selection between groups.

However, theoretical conditions can be imagined in which group selection wins out over individual selection. Darwin's suggestion is not necessarily wrong, or incoherent. But since he wrote, biologists have become much more concerned about the exact conditions needed for group selection to operate. Darwin's arguments would be much more controversial among his modern followers than his innocent wording allows for. Nevertheless, it is striking how Darwin has identified the conflict between individual and group advantages in the evolution of 'the social and moral faculties'. His suggestion that they evolved by group selection remains possible, even if it has become controversial.

Biologists since Darwin have added one further factor that is not in his list here. That factor is usually called kin selection. Natural selection can favour self-sacrifice if it benefits the genetic relatives of the individual who is sacrificing himself or herself. The genes in an individual are, to a certain probabilistic extent, also present in his or her siblings and cousins. A gene has a fifty-fifty chance of being shared between full siblings. Thus, if an individual sacrifices his own life but in such a way as to more than double the eventual reproduction of his siblings, natural selection favours self-sacrifice.

Darwin almost certainly never thought of kin selection. There is a passage in *The Origin of Species* about sterile castes in ants that once was thought to hint at it, but a careful reading of that passage shows that it was concerned with another topic completely. The major publications on kin selection did not come until 1964, and were written by W. D. Hamilton. In all, biologists still recognize self-sacrificial behaviour as a major challenge for Darwin's theory of natural selection. They now discuss four possible solutions: kin selection, reciprocal altruism, group selection and cultural factors. In a cultural explanation, self-sacrifice is not favoured by natural selection; it exists because of cultural factors of one kind or another, such as Darwin's 'praise and blame'. Of these four explanations, Darwin's discussion contains at least hints of three of them; only kin selection is clearly a post-Darwinian discovery. Modern biologists differ from Darwin in that they do not base reciprocity on rational calculation and foresight, and they are more sceptical about the power of group selection. Despite these differences, Darwin's analysis reads, conceptually and despite its Victorian language, very like a modern one.

8

NATURAL SELECTION AS AFFECTING CIVILIZED NATIONS

The Descent of Man 2

Some remarks on the section of natural selection on civilised nations may be worth adding. ... With savages, the weak in body or mind are soon eliminated; and those that survive commonly exhibit a vigorous state of health. We civilised men, on the other hand, do our utmost to check the process of elimination; we build asylums for the imbecile, the maimed, and the sick; we institute poor-laws; and our medical men exert their utmost skill to save the life of every one to the last moment. There is reason to believe that vaccination has preserved thousands, who from a weak constitution would formerly have succumbed to small-pox. Thus the weak members of civilised societies propagate their kind. No one who has attended to the breeding of domestic animals will doubt that this must be highly injurious to the race of man. It is surprising how soon a want of care, or care wrongly directed leads to the degeneration of a domestic race; but excepting in the case of man himself, hardly any one is so ignorant as to allow his worst animals to breed.

The aid which we feel impelled to give to the helpless is

mainly an incidental result of the instinct of sympathy. [. . .] Nor could we check our sympathy, even at the urging of hard reason, without deterioration in the noblest part of our nature. The surgeon may harden himself whilst performing an operation, for he knows that he is acting for the good of his patient; but if we were intentionally to neglect the weak and helpless, it could only be for a contingent benefit, with an overwhelming present evil. We must therefore bear the undoubtedly bad effects of the weak surviving and propagating their kind; but there appears to be at least one check in steady action, namely that the weaker and inferior members of society do not marry so freely as the sound. [. . .] It was established from an enormous body of statistics, taken during 1853, that the unmarried men throughout France, between the ages of twenty and eighty, die in a much larger proportion than the married: for instance, out of every 1,000 unmarried men, between the ages of twenty and thirty, 11.3 annually died, whilst of the married only 6.5 died. [. . .] Dr Stark considers that the lessened mortality is the direct result of 'marriage, and the more regular domestic habits which attend that state'. He admits, however, that the intemperate, profligate, and criminal classes, whose duration of life is low, do not commonly marry; and it must likewise be admitted that men with a weak constitution, ill health, or any great infirmity in body or mind, will often not wish to marry, or will be rejected. [. . .] On the whole we may conclude with Dr Farr that the lesser mortality of married than of unmarried men, which seems to be a general law, 'is mainly due to the constant elimination of imperfect types, and to the skilful selection of the finest individuals out of each successive generation'.

Most of the part of *The Descent of Man* that is about human evolution is concerned with evolution in the past. Darwin did not have the chronology that is now available to us, but his book was mainly concerned with events in the human line from about 5 million years ago to about 25,000 years ago. During this time, our ancestors evolved several differences from other apes, in such features as large brains and an upright, bipedal posture. By about 25,000 years ago (plus or minus a bit, depending on the region of the globe) humans who were indistinguishable from us were well established. However, he also added a section on 'natural selection as affecting civilised nations'. Darwin and his contemporary readers may have been more confident about the meaning of 'civilized' than would be most readers today, but his initial sentences make clear what is crucial for his argument. Natural selection acts by differences in death rates, as some individuals die and others survive, and by differences in fertility, as some survivors produce more offspring than others. By 'civilised nations' Darwin means societies that have medical, health and welfare systems that 'check the process of elimination'. Vaccination, for instance, keeps people alive who would otherwise die of infectious disease. It is therefore possible that the action of natural selection is being slowed down, or even prevented, in certain societies.

In the opening sentences, modern readers need to allow for changes in language use. Expressions such as 'asylums for the imbecile' or 'degeneration of a domestic race' clearly allude to sensitive topics which attract what linguists call euphemism creep, as people introduce new words that initially lack the negative connotations of existing words (such as 'imbecile'), but those new words then acquire much the same negative associations, leading to the introduction of yet newer words.

No doubt a discussion of the same topic today will appear as insensitive to readers in 135 years' time as does Darwin's to some readers now. What is distinctive about Darwin is not his use of language, but his power of argument. He is one of the greatest thinkers of all time, and most Darwin readers will want to follow, understand and be stimulated by his arguments rather than sidetracked by his language.

Arguments about whether natural selection is relaxed in some human societies have been controversial ever since Darwin's time. What is (again) striking is the way Darwin identified essentially all the main themes of subsequent discussions. Indeed, he identified rather more than all the themes of many writers of the twentieth and twenty-first centuries for and against eugenics.

Darwin begins by suggesting that natural selection may be relaxed in 'civilised nations'. By analogy with domestic animals, this can be expected to lead to a deterioration in human quality over the generations because inferior types have not been eliminated. One solution might be to return to natural selection, putting a stop to medical intervention. Darwin rejects this option for ethical reasons. It would represent 'a deterioration in the noblest part of our nature'. And 'if we were intentionally to neglect the weak and helpless, it could only be for a contingent benefit, with an overwhelming present evil'. This passage is worth noting, because Darwin is sometimes charged with being a eugenicist – by authors who have perhaps only read the previous sentences about 'the degeneration of a domestic race'. However, it pays to read on. Darwin immediately rejects any return to natural selection. It would mean an 'overwhelming present evil', in the suffering of people left to die in the absence of medicine and welfare support. Indeed it is clear he is not even sure that civilization has created the conditions for degeneration. He says that neglecting the weak and helpless 'could only be for a contingent benefit'. That is, it might prevent degeneration – or it might not. It all depends. If natural selection is indeed relaxed, degeneration is likely to follow. But is selection relaxed in humans? Darwin goes on to look at mortality rates in married and unmarried people. There was already extensive evidence that unmarried humans die off at a higher rate than equivalent married humans. In the data known to Darwin, unmarried men die at about twice the rate of married men of the same age and living in the same place.

Darwin considers two explanations for the difference in mortality rate. One is that marriage itself may cause a reduction in mortality, for instance if society discriminates in favour of married over single individuals. The other is that healthier individuals may have an advantage in the marriage market, and the single individuals are the less healthy leftovers. In that case, the difference in mortality rates between married and unmarried individuals is not caused by change in quality of life following marriage. The difference arises because the marriage market sorts individuals according to quality. Darwin concludes in favour of the second explanation. On this conclusion, the idea that selection is relaxed in humans may be invalid. We may have relaxed some forms of selection, by vaccination and surgery, but other forms remain, in the way we pick our marriage partners. Civilized humans may not be on the road to degeneration at all.

In modern human biology, the issues raised by Darwin are still live. Darwin's first point was that if natural selection is relaxed, the population will decline in quality over time. The decline occurs because inferior genes – biologists call them deleterious mutations – are normally eliminated by natural selection. Individuals that contain poor quality genes are more likely to die before reproduction, and that removes the inferior genes from the population. However, there is a steady input of new harmful genes as they arise by mutation every generation. In most populations of living creatures, there is an approximate balance between the input of new mutations and their removal by natural selection. In a population in which harmful mutant genes are not removed by selection at the same rate as they arise, the quality of its members necessarily declines over the generations. Biologists do experiments on the process. If natural selection is prevented from acting on fruitflies (a standard lab animal), the life expectancy of the flies goes down from one generation to the next. To be exact, the ability of the flies to survive decreases by about half a per cent per generation. After thirty generations without selection, the viability of the experimental flies is down to about 85% of its initial figure.

The rate of decline in quality in a population in which natural selection is not acting depends on the mutation rate. If mutations occur at a high rate, the population decays rapidly; if they occur at a low rate, the decay is slower. Biologists currently disagree about what the mutation rate is. If natural selection has truly stopped acting on humans in some countries, we can be sure their DNA will be randomized over time. However, we do not know whether the randomization would become significant after a few, or tens, or hundreds of generations. In any case, Darwin's basic claim – that life decays if natural selection stops acting – is still accepted.

Darwin's second point was an ethical one – that we are right to use medicine. It may interfere with natural selection, but that is just too bad. Modern opinion would, I suspect, agree even more strongly with Darwin than would that of his contemporaries. Between then and now, certain (though not all) eugenicists argued that we should either restore natural selection against poor quality genes, or use technology to mimic natural selection – for example, by sterilizing people judged to be genetically unfit. Legislation to do just that was enacted in several countries, and eugenic policies were implemented. Nazi Germany was one such country, but not the only one; indeed the Nazis copied their legislation from laws in the USA. From the mid-twentieth century on, eugenic policies became politically unpopular and eugenic legislation was repealed. Modern societies, like Darwin, are willing to put up with any conjectural future genetic decay, whether the alternative is eugenic law or a return to natural selection. Some minority voices disagree; but they are minority voices, and well aware of the fact.

Finally, Darwin challenged whether selection really is relaxed in any human societies. It may be that medicine has relaxed some forces of selection, but the marriage market may still work against bad genes. Subsequent research has amply supported Darwin in his factual point about the mortality of married and unmarried people. Large surveys, up to about the 1960s, document the difference beyond doubt. The difference is found in all countries that have been looked at, and in both men and women. On average, unmarried men are about 1.8 times more likely to die than equivalent married men. The difference is more like 1.5 times for unmarried relative to married women. After about 1970 or so, the statistics become less interesting. Reproduction outside wedlock became commoner in many countries, as has marriage without reproduction. For the workings of natural selection, what matters is any difference in genetic quality between people who breed and people who do not breed. In Darwin's day, and for some decades after him, the relative mortalities of married and unmarried people was a rough and ready way of studying that question. It would be much harder to study it in the current population of, for example, the USA or any European country. We should need to know the relative mortalities of people who became fathers, as compared with men who do not become fathers; and of people who become mothers, as compared with women who do not become mothers.

Although the mortality difference between unmarried and married people has become better documented in the century since Darwin, its explanation has not become any clearer. Biologists and social scientists have continued to debate between the two explanations considered by Darwin. Either marriage makes you less likely to die, or being less likely to die makes you more likely to marry. Darwin, we saw, thought the second factor more important. But his argument (incompletely quoted here) is too brief to be convincing. It is practically impossible to obtain decisive evidence for humans. For non-humans, biologists have found good evidence from several species that individuals with superior genes are more successful in the mating market. This lends some support to Darwin's interpretation, but probably not enough to convince a sceptic. It could still be that, in humans, the mortality difference between married and unmarried people is an effect of being married – or, nowadays, of being pair-bonded.

One other aspect of Darwin's argument is worth noticing: it assumes that the mortality difference between married and unmarried people has some genetic component. Natural selection can only be working via the marriage market if people with inferior genes are failing to marry. It could be that people with poor health are less likely to marry, but that health differences are entirely due to non-genetic factors. Then the marriage market would discriminate against unhealthy people, but not against bad genes. It is a reasonable assumption that genes have some influence on health and mortality rates – we have plenty of evidence of genetic disease, for instance – but it is still an assumption. It is an assumption that would be difficult to test. We cannot do the kind of experiments that would be needed to show that the mortality difference between married and unmarried people has a genetic component.

Biologists have also identified other ways that natural selection could be acting in humans. Another possibility, besides the marriage market, is selection early in the life cycle, including selection among sperm and selection among eggs. Much of medicine is concerned with the elderly, and may make little difference to how selection acts on a population. Elderly people are past reproductive age. If medicine keeps an old person alive for an extra ten years, that has no effect on the genetic composition of the population in the next generation. Medicine matters (in this discussion) when it keeps alive someone who then reproduces, and we can therefore ignore all medical activity in people of post-reproductive age.

Much of the natural selection against bad genes may occur much earlier in the life cycle. Women produce millions of the cells that can develop into eggs, but only a few dozen of them ever become eggs. Of the ones that do develop into eggs, only a few are fertilized and begin development. Only about 30% of conceptions (in biological terms, zygotes) make it to term as a baby, the other 70% perish. Men produce billions of sperm, but at most only a tiny minority ever succeed in carrying their DNA into the next generation. There is a massive mortality in sperm and eggs, and in the early stages of embryonic development. We do not know what fraction of this mortality is due to natural selection, but we do know that some of it is. It could be that natural selection mainly acts to remove bad genes from the human population by the failure of some gametes to make it into zygotes and the failure of some zygotes to develop into babies. Medicine has done almost nothing to reduce this kind of early mortality. If medicine has relaxed selection, it is in later stages – between birth (or slightly before) and old age. Thus, natural selection could be acting in 'civilised nations' in much the same way as it has always acted during human evolution.

In all, the topic of whether natural selection is relaxed, or even suspended, in humans by medicine and welfare, remains of pressing interest. Some authors assume that selection is relaxed in humans. They may be right, and humans (in some countries) may be embarked on a unique evolutionary journey, as their DNA is randomized over time. Civilization would be leading us into extinction. However, it is not certain that selection is relaxed in humans. As Darwin notes, selection could still act via the mating market. There is the further possibility that selection acts in the early stages of the life cycle. Civilization could then express the 'noblest part of our natures' but without relaxing the force of natural selection or toppling us into genetic decay.

9

SEXUAL SELECTION

The Descent of Man 3

With animals which have their sexes separated, the males necessarily differ from the females in their organs of reproduction; and these are the primary sexual characters. [. . .] There are, however, other sexual differences quite unconnected with the primary reproductive organs, and it is with these that we are more especially concerned – such as the greater size, strength, and pugnacity of the male, his weapons of offence or means of difference against rivals, his gaudy colouring and various ornaments, his power of song, and other such characters.

We are here concerned with sexual selection. This depends on the advantage which certain individuals have over others of the same sex and species, solely in respect of reproduction. [. . .] There are many structures and instincts which must have been developed through sexual selection – such as the weapons of offence and the means of defence of the males for fighting with and driving away their rivals – their courage and pugnacity – their various ornaments – their contrivances for producing vocal or instrumental music – and their glands for emitting odours, most of these latter structures serving only to allure or excite the female. It is clear that these characters are the result of sexual and not of ordinary selection, since unarmed, unornamented, or unattractive males would succeed equally well in the battle for life and in leaving a numerous progeny, but for the presence of better endowed males. We may infer that this would be the case, because the females, which are unarmed and unornamented, are able to survive and procreate their kind. Secondary sexual characters of the kind just referred to will be fully discussed in the following chapters, as being in many respects interesting, but especially as depending on the will, choice and rivalry of the individuals of either sex. When we behold two males fighting for the possession of the female, or several male birds displaying their gorgeous plumage, and performing strange antics before an assembled body of females, we cannot doubt that, though led by instinct, they know what they are about, and consciously exert their mental and bodily powers.

Just as man can improve the breed of his game-cocks by the selection of those birds which are victorious in the cockpit, so it appears that the strongest and most vigorous males, or those provided with the best weapons, have prevailed under nature, and have led to the improvement of the natural breed or species. A slight degree of variability leading to some advantage, however slight, in reiterated deadly contests would suffice for the work of sexual selection; and it is certain that secondary sexual characters are eminently variable. Just as man can give beauty, according to his standard of taste, to his male poultry, or more strictly can modify the beauty originally acquired by the parent species, can give to the Sebright bantam a new and elegant plumage, an erect and peculiar carriage – so it appears that female birds in a state of nature, have by a long selection of the more attractive males, added to their beauty or other attractive qualities. No doubt this implies powers of discrimination and taste on the part of the female which will at first appear extremely improbable; but by the facts to be adduced hereafter, I hope to be able to shew that the females actually have these powers.

The majority of *The Descent of Man* is about Darwin's theory of sex differences, a theory he called (as it is still called) sexual selection. The question that the theory is designed to answer is, why is it that males in many species have evolved apparently disadvantageous attributes – attributes that reduce the chance of survival? The peacock's tail proved a dramatic example of the problem. (The peacock's 'tail' is, strictly speaking, a development of the back feathers not the tail feathers; but it is convenient to refer to it by its common name.) The tail is a large and extravagant ornament. It is expensive to grow; its bright colour attracts predators and its size reduces flight efficiency. A peacock would survive better without it – and yet it has evolved somehow.

The peacock's tail is an apparently maladaptive attribute. We saw, in Chapter 1, that Darwin's first test of his (or anyone else's) theory of evolution is whether it explains adaptation. Life is full of examples of adaptation; Darwin often used the woodpecker's beak as an example, but we could now add examples from molecular biology to social behaviour. Natural selection passes Darwin's first test because it readily explains adaptation.

But the success of natural selection in explaining adaptation can be turned against it: some attributes of living things seem

not to be adaptive, and that in turn suggests the theory of natural selection may be wrong or incomplete in some way. If
natural selection were all-powerful, things like the peacock's
tail should not exist. It was because these extravagant sexual
attributes posed such a deep challenge to his theory that
Darwin spent so much time thinking and gathering evidence
about them. The 500 or so pages of Part II of *The Descent of
Man* are the fruits of that work.

Darwin begins by specifying more exactly the kinds of
attributes that he is concerned with. He distinguishes 'primary' from 'secondary' sexual characters. (Darwin uses the
word 'character' in the standard biological way. 'Character'
here does not mean personality, but refers to any attribute, or
property, of an organism.) The primary sexual characters are
the organs of reproduction – genitalia, ovaries and testes. It is
unremarkable that these differ between the sexes, given the
fact of sexual reproduction. They have been shaped, during
evolution, by ordinary natural selection. Secondary sexual
characters are organs that are not simply needed for reproduction but that differ between the sexes and that seem to be
used in some way during reproduction. The peacock's tail is a
secondary sexual character.

However, not all secondary sexual differences pose a puzzle
for the theory of natural selection. In a passage that I have not
included in the quotation, Darwin discusses various kinds of
claspers that are found particularly in the males of aquatic
animals, such as some kinds of crustaceans (the group that
includes shrimp and crabs). These are structures used by the
male to grip the female. They may be needed to prevent the
male and female from being separated by water currents
before copulation is complete. The form of the claspers is
probably shaped by natural selection. There are yet other sex

differences that make good sense on the theory of natural selection and that Darwin is also not concerned with. He discusses (again, not in the quoted passage) how the beaks of the males and females differ in certain bird species. This is probably because the two sexes have different 'ways of life'; they may eat different foods, for instance. The male and female beaks could have been shaped by natural selection, such that individuals of each sex are optimally efficient at feeding. But simple natural selection does not explain all sex differences. We are left with the puzzle of 'the greater pugnacity of the male, his weapons of offence', 'his gaudy colouring and ornaments, his power of song' and so on.

Darwin then goes on to argue that these secondary sexual characters are not due to ordinary natural selection. His reasoning is based on the form of the females in the species in question. If some male attribute, such as antlers or brightly coloured feathers, were needed for survival then it should be found in the females too. Its absence from females suggests (particularly after we have looked some more at the theory of sexual selection) that the optimal state for members of the species is to lack the antlers, or feather ornaments. Males would survive better without them. But the peculiar forces of sexual selection have caused them to evolve, reducing the efficiency of males in the non-reproductive parts of life.

What is sexual selection? Darwin distinguishes two main kinds, which are now referred to as male competition and female choice. Males may fight each other over females. The males that are stronger, or have superior weaponry, will then be more likely to reproduce. Over the generations, males will evolve ever more powerful weaponry. The weaponry could be net advantageous even if it was so cumbrous that it reduced the male's survival rate. In evolutionary terms, lowered survival can be traded off against superior reproduction. If larger weapons halve a male's chance of survival but triple his chance of breeding, they will evolve. Competition between males can therefore explain some apparently maladaptive secondary sexual characters.

Organs used in fights between males fit Darwin's description of sexual selection. He says that sexual selection works by the advantage that some individuals have relative to other individuals of the same sex and species, during reproduction. In Chapter 2, we saw how it was distinctive of Darwin's thinking that he saw competition (the struggle for existence, in his words) as between individuals within a species. His theory of sexual selection not only illustrates the same theme but takes it a step further. Competition is not just within a species. If food resources are in short supply, an individual will typically compete against all members of its own species for food, and perhaps some members of closely related species. This competition – to stay alive – is not in most cases much influenced by gender. But competition to reproduce is highly gendered. Males do not compete with females to settle who succeed in producing offspring; males compete only with males.

In genetic terms (which Darwin could not have used) we can say that, of all the genes sent by one generation into the next generation, half the genes are sent on by males and half by females. No male act can substitute male genes for female genes during reproduction. If a male is a strong fighter, he can increase his share of the male genes sent into the generation; but no amount of fighting will enable him to take any of the female share. In this sense, reproductive competition is confined within each gender within each species. Darwin differed from most of his contemporaries, when he suggested that competition is between individuals within a species, rather than between species or even between a species and inanimate nature. But in his theory of sexual selection, competition takes place not just within a species, but within a gender within a species.

Darwin's second mechanism of sexual selection is female choice. Competition between males can explain male secondary sexual organs that function in fighting. However, males in some species also have ornaments, such as the peacock's tail, that would be worse than useless in a fight. For these, Darwin put forward a more audacious hypothesis; just as humans have artificially bred ornamental poultry, so 'female birds in a state of nature, have by a long selection of the more attractive males, added to their beauty'. Naturalists before Darwin had recognized that males fight over females, but no one had suggested anything quite like Darwin's theory of female choice before.

Similarly, after Darwin, biologists generally accepted that some male attributes, such as strength and weaponry, are due to competition between males. But his hypothesis of female choice has proved more controversial. One reason is that Darwin expressed his idea in terms of conscious aesthetic choice – a topic I'll return to below. Another reason is that it is unclear in Darwinian terms why females should have evolved to choose in the way Darwin suggests. If peahens do in fact mate preferentially with peacocks who have larger or brighter tails, then that helps explain why males have these structures. In males, the reduced survival that results from having an extravagant tail is traded off against superior reproductive success. The structure will evolve, by the standard Darwinian mechanism.

But the argument immediately raises the next question:

why has natural selection favoured females who choose mates that have huge, colourful, survival-reducing tails? For this choice to have evolved, females who were choosy had to leave more offspring than females who mated indiscriminately. For choice to be maintained now, females still need to be gaining some advantage from it. Darwin did not discuss this problem. In a way, he did not have to. If females choose colourful males, males will evolve to be colourful. Colourfulness has then been (conditionally) explained. For any explanation, you can always ask a 'why' question that takes the explanatory problem one step back – and it is not a defect in a theory if it stops its explanatory work at a certain stage.

However, in the case of female choice and extravagant, costly structures such as the peacock's tail, the question is acute. Darwin had no evidence for female choice. Indeed it is only recently (in the 1990s) that biologists have shown that peahens do mate preferentially with bigger and brighter tailed males. Moreover, the choice hypothesis by Darwin is quite paradoxical. A female chooses a male that has an attribute that reduces survival. If so, the attribute will be inherited by her sons, and reduce their survival. Therefore, if a female picked a less extravagantly ornamental mate, her sons would survive better and her reproductive output would be increased. Natural selection seems to work against the female choice Darwin hypothesized.

Biologists have been fascinated by this problem – the evolution of female choice – for almost a century. One solution was first proposed by R. A. Fisher in 1916. He argued that female choice could evolve in a 'runaway' process, resulting in the choice of over-ornamented males. Once all the females in a population choose in a certain way, the majority preference acts as a kind of trap that none can escape. If one female chooses a less ornamented mate, her sons will indeed have a higher survival – but when they grow up they will be in a population where most females discriminate against unornamented males. Each female has to pick an ornamented mate, in order to produce sons who will later be successful at mating.

Fisher's idea has been much discussed. Some biologists still support it, others reject it, many are uncertain whether it is right or wrong. Other ideas suggest that male ornaments reveal desirable qualities – such as genetic quality, or resistance to disease. A female who picks an ornamented mate can then produce healthy offspring, of high genetic quality like their father. These ideas also remain controversial. In all, female choice is still, as Darwin suggested, the best explanation we have for 'ornamental' male attributes that do not function in competition between males. We now have evidence, which Darwin lacked, that females do in fact discriminate in mating, favouring certain kinds of males. However, the puzzle of why females in some species seem to prefer extravagantly ornamented mates – the puzzle of the evolution of female choice – has not been solved. Good ideas have been proposed and studied, but none commands wide acceptance among biologists.

The other controversial feature of Darwin's theory is that he expressed it in terms of conscious mental powers. 'When we behold two males . . . performing strange antics before an assembled body of females, we cannot doubt that . . . they know what they are about, and consciously exert their mental and bodily powers.' Female choice, likewise, Darwin thought to be conscious, based on 'powers of discrimination and taste'. When Darwin wrote 'we cannot doubt that', he was saying something that biologists and psychologists soon came to doubt very strongly indeed. Around the beginning of the twentieth century, a science of behaviour was starting to be founded. It was, practically, based on a rejection of ideas like Darwin's, about animal consciousness. Scientists came to realize that apparently complex behaviour, such as choosing a mate, can be produced by simple mechanisms. 'Higher' mental powers, such as conscious reasoning, were rejected from the new twentieth-century science of behaviour.

The rejection could take two forms. For some, it was methodological. Consciousness is impossible to study scientifically in animals. Therefore, for scientific purposes, we ignore it. We study other aspects of behaviour for which scientific methods can be used. Maybe non-humans use conscious reasoning, maybe they do not – but we do not need to answer that question in order to make various kinds of scientific progress. Others took a harder line, and argued that consciousness is unique to humans. Either way, Darwin's wording looked retrospectively unfortunate. None of his main arguments depend on consciousness in non-human animals. He clearly thought that birds, and other life forms, used conscious striving like ours. But if they do not, Darwin's main points hold up. Whether displaying males consciously or unconsciously strive to impress females and out-compete other males, the evolutionary result will be the kinds of male organs we see and wish to explain. Whether females choose their mates by conscious aesthetic judgements or unconscious decision-mechanisms, if they do choose somehow it will explain certain male attributes. Thus, in one respect, our modern understanding of sexual behaviour differs from Darwin's. For Darwin, mating in many animals, including birds and perhaps even insects as well as humans, was a world of conscious rivalry and aesthetic choice. For most modern thinkers, male displays and female choice are robotic. Darwin's statements about consciousness have dropped out of modern accounts of sexual selection, though his theory is still used to explain sex differences, including such bizarre attributes as peacocks' tails. In that sense, Darwin invented a highly successful theory. But the way the theory is used nowadays ignores one factor — consciousness — that was crucial for Darwin.

10

THE EXPRESSION OF THE EMOTIONS

No doubt as long as man and all other animals are viewed as independent creations, an effectual stop is put to our natural desire to investigate as far as possible the causes of Expression. By this doctrine, anything and everything can be equally well explained; and it has proved as pernicious with respect to Expression as to every other branch of natural history. With mankind some expressions, such as the bristling of the hair under the influence of extreme terror, or the uncovering of the teeth under that of furious rage, can hardly be understood, except on the belief that man once existed in a much lower and animal-like condition. The community of certain expressions in distinct though allied species, as in the movements of the same facial muscles during laughter by man and by various monkeys, is rendered somewhat more intelligible, if we believe in their descent from a common progenitor. He who admits on general grounds that the structure and habits of all animals have been gradually evolved, will look at the whole subject of Expression in a new and interesting light. [. . .]

I will begin by giving the three Principles, which appear

to me to account for most of the expressions and gestures involuntarily used by man and the lower animals, under the influence of various emotions and sensations. [. . .]

I. *The principle of serviceable associated Habits.* Certain complex actions are of direct or indirect service under certain states of the mind, in order to relieve or gratify certain sensations, desires, &c.; and whenever the same state of mind is induced, however feebly, there is a tendency through the force of habit and association for the same movements to be performed, though they may not then be of the least use. Some actions ordinarily associated through habit with certain states of mind may be partially repressed through the will, and in such cases the muscles which are least under the separate control of the will are the most liable still to act, causing movements which we recognise as expressive.

II. *The principle of Antithesis.* Certain states of the mind lead to certain habitual actions, which are of service, as under our first principle. Now when a directly opposite state of mind is induced, there is a strong and involuntary tendency to the performance of movements of a directly opposite nature, though these are of no use; and such movements are in some cases highly expressive.

III. *The principle of actions due to the constitution of the Nervous System, independently from the first of the Will, and independently to a certain extent of Habit.* When the sensorium is strongly excited, nerve-force is generated in excess, and is transmitted in certain definite directions, depending on the connection of the nerve-cells, and partly on habit: or the supply of nerve-force may, as it appears, be interrupted. Effects are thus produced which we recognise as expressive. This third principle may, for the sake of brevity, be called that of the direct action of the nervous system.

For Darwin, writing *The Descent of Man* (published in 1871) and then *The Expression of the Emotions* (published in 1872) was one long, continuous act. As soon as he had finished checking the proofs of *The Descent of Man*, he started writing *The Expression of the Emotions.* Darwin was now 63 years old and his health had been weak for years: it is not surprising that he was completely exhausted. However, he recovered to write several further books in his final decade.

The Descent of Man and *The Expression of the Emotions* are closely related. Indeed Darwin probably originally planned them as a single work. They were split partly for reasons of length, but also because the theoretical system underlying *The Expression of the Emotions* took on a life of its own. In *The Descent of Man*, we saw (in Chapter 7 above) how Darwin looked at a series of human social and mental faculties, such as language, morality, religion, social cooperation and self-sacrifice. These mattered because creationists, who thought that humans had separate origins from the rest of life, tended to argue that the social and mental faculties were what differentiated humans from the rest of life. Our bodies may look like a modified form of an ape's body, but (the argument ran) apes have nothing like our moral, social and religious senses. Darwin replied by showing how all these faculties had evolved in humans, by natural selection, from precursors in non-human forms of life.

A similar stimulus had originally inspired Darwin's research on the emotions. In 1838 he read a book on the subject by an expert, Sir Charles Bell. Bell said that there were certain facial muscles that existed uniquely in humans and that functioned in emotional expression. Likewise, it was a common argument, among eighteenth-century moral and political thinkers, that humans had uniquely been equipped with blushing. Human social life would be impossible if people told each other lies all the time, and blushing helps to prevent, or reduce the social damage done by, lying. The expression of the emotions was held to be another attribute, like language and morality, that distinguished humans from the rest of life.

Darwin set to work, gathering information for over thirty years before he wrote his book. He already knew in 1838 that Bell's claims about facial muscles were in error. All the same facial muscles are present in humans and non-human apes. Darwin's copy of Bell's book still survives, along with Darwin's marginal comments. At one point, Bell discusses the muscle (the corrugator supercilii, to be exact) that knits the eyebrows and 'unaccountably but irresistibly conveys the idea of the mind'. Darwin comments: 'Monkey here? . . . I have seen well developed in monkeys . . . I suspect he never dissected monkey.' Darwin went on to trace continuities between the forms of emotional expression in non-human and in human life. It turns out there is nothing uniquely human about emotional expression.

The opening paragraph of the quotation takes the matter further. There are general objections to any non-evolutionary view of life, but specific problems also arise for the emotions. Our teeth are reduced from those of apes and most monkeys. In male baboons, canine teeth are like daggers – they are dangerous and even deadly weapons that are displayed in conflicts and used in fights. Chimpanzees also have large canines, and use them in fights. It makes sense, therefore, that our primate ancestors flashed their canines – bared their teeth – as a threat display, expressing the emotion of anger or aggression. Humans continue to bare their teeth in threats, or in sneers, even though our teeth are insignificant as weapons. As Darwin says, baring the teeth while in a rage 'can hardly be understood' except on the theory of evolution.

One purpose of *The Expression of the Emotions* continued to be to trace the evolutionary continuity of emotional expression between non-human and human life. However, that purpose came to be secondary, as Darwin devised a whole theoretical system to understand emotional expression. The main question that underlies the book is why emotional expression takes the form it does. Why do we express good spirits by smiling and laughter? Sadness by weeping, grief by knitted eyebrows, wrinkled forehead and a turned-down mouth? Why do we shrug our shoulders when helpless? The main chapters of the book work through the emotional states, describing the form of their expression and thinking about why expression takes the form it does. The book is a fascinating read, not least because of the range of material Darwin had gathered. He observed his fellow adults, and himself, but he also particularly observed his children. The birth of his first children in the 1840s was a further great stimulus to the work. He looked at emotional expression in painting and sculpture, and how it was described in literature. He collaborated in work in which facial muscles were electrically stimulated, to see how those muscles affected the form of the face. He sent out questionnaires to people living in all corners of the globe, asking them how the local inhabitants expressed their emotions. The answers led Darwin to conclude that most forms of emotional expression are universal in humans. The conclusion

proved controversial among some twentieth-century anthro-
pologists, but it is now widely accepted, particularly following
the research of Paul Ekman. The contents of *The Expression of
the Emotions* are mainly non-technical. They also have a per-
sonal interest for the (human) reader, making it one of
Darwin's most readable books.

Darwin begins the book by setting out his general explanatory system. It consists of three principles, and the second half
of the quoted passage sets them out. Darwin uses these principles, often not explicitly, in the main chapters of the book –
the chapters about particular emotions. The three principles
do not amount to as strong a theoretical system as does, for
instance, the theory of natural selection in *The Origin of
Species*. Natural selection provided a powerful explanation for
the facts he had gathered about species. In *The Origin of Species*
he continually refers to, or applies, the theory to make sense
of one fact or another. In contrast, the three principles of
emotional expression made good sense of some facts, but
were less immediately applicable to others. In the book,
Darwin seems to draw on the principle at some points but
then ignore them for long passages elsewhere. The book is not
one long argument in favour of the three principles. It is
more a set of reflections, with a preliminary theory. Darwin
liked to have a general idea to organize a subject; but in the
case of the emotions, his reflections on individual topics
sometimes took over from any need to test his general theory.

Nevertheless, the three principles are the central theory of
the book and are never far behind what Darwin has to say.
They are therefore well worth understanding. The first principle in the quoted passage is what Darwin calls 'serviceable
associated habits'. Some actions require us to adopt a particular posture before or during the action. For instance, in a